KW-266-154

THE VALUE OF
AGRICULTURAL LAND

THE VALUE OF

AGRICULTURAL LAND

By

Colin Clark, M.A., D. Litt.

Monash University, Australia

Formerly Director,
Agricultural Economics Institute,
Oxford

WITHDRAWN FROM UNIVERSITY OF PLYMOUTH LIBRARY SERVICES

SEALE-HAYNE NEWTON ABBOT DEVON AGRICULTURAL COLLEGE

PERGAMON PRESS

OXFORD · NEW YORK · TORONTO
SYDNEY · BRAUNSCHWEIG

Pergamon Press Ltd., Headington Hill Hall, Oxford

Pergamon Press Inc., Maxwell House, Fairview Park, Elmsford,
New York 10523

Pergamon of Canada Ltd., 207 Queen's Quay West, Toronto 1

Pergamon Press (Aust.) Pty. Ltd., 19a Boundary Street,
Rushcutters Bay, N.S.W. 2011, Australia

Vieweg & Sohn GmbH, Burgplatz 1, Braunschweig

Copyright © 1973 Dr. C. Clark

*All Rights Reserved. No part of this publication may be
reproduced, stored in a retrieval system, or transmitted, in any
form or by any means, electronic, mechanical, photocopying,
recording or otherwise, without the prior permission of
Pergamon Press Ltd.*

First edition 1973

Library of Congress Cataloging in Publication Data

Clark, Colin, 1905–
   The value of agricultural land.

   Includes bibliographical references.
   1. Farms--Valuation.   I.   Title.
HD1393.C49   1973   333.3'35   73-179
ISBN 0-08-017070-6

Printed in Hungary

CONTENTS

PREFACE

CERTAIN monosyllabic words have the emotional impact of a bullet, with the power to kill dead any rational discussion of the economics of the subject—and politicians, whose stock-in-trade is words, know well how to use them. One of these words is *land*.[1] When this word is spoken it immediately conjures up visions: to some, of hard-working, down-trodden farmers, paying extortionate rents and subject to humiliating social oppression; to others, of a peaceful rural civilisation presided over by humane and gracious landowners, fully aware of their economic and social responsibilities. From this conflict of ideas bloody wars are generated, from the peasant revolts of the Middle Ages to Spain and Vietnam in our time, and perhaps South America and India in the near future.

But the existence of these terrible and overriding truths in no way absolves the economist from his duty of analysing the circumstances under which land is valuable, or less valuable, in the sense that people are or are not willing to pay high rents and prices for it. Whether land should be owned by the state, or by public corporations, or by "collectives", or by large or by small private landowners, or exclusively by those who cultivate it, is a political question on which the economist, as such, cannot claim the right of decision. Political changes in the forms of land ownership may be very desirable. But the economist is entitled to point out that they will not (other circumstances remaining the same) alter the value of agricultural land, or the amount of "economic rent" which will eventually accrue to its (public or private) owner. The removals of former landowners in Russia, China and Vietnam, and

---

[1] The others are *gold*, *steel*, *wheat* and *oil*.

of the *Samurai* in Japan, were followed by similarly burdensome taxes or compulsory deliveries. What will reduce, substantially and permanently, the proportion of the product of agriculture payable as rent to the owner of the land, as is shown below by several interesting examples, is the increasing availability of non-agricultural employment.

When agriculture is subsidised, or given tax concessions, or helped in any other way, the benefit accrues in the form of a rise in the price of land, rather than in the income of the farmer as such. Where (as is now the case in most countries) the farmer does own his land, it is second nature to him to think that land prices ought to go up. For the sake of attracting new enterprise into farming, for securing a rational reorganisation of farms to meet changing circumstances, and for the public interest generally, it is desirable (except in a few remote areas) that land prices should come down.

CHAPTER 1

GENERAL PRINCIPLES

THE market for agricultural land differs from most other markets in that the supply is virtually fixed. Under such circumstances, it is often said, price will be determined by demand alone. This is not a very accurate way of putting it. The precise statement of the situation would be that the stock being unchangeable, the price will have to adjust itself to such a level that those who inescapably will have (collectively) to hold this stock will, in fact, be content so to hold it. When we state the situation in this manner, we see that there are a number of other markets besides that for land to which the same considerations apply —the difference being, of course, that in the case of agricultural land the supply is fixed permanently, whereas in other markets it is fixed for a shorter period. Examples include any agricultural product for which there is a single harvest season; once the harvest is gathered, supply can on no account be augmented until the next harvest, and until then therefore price has to adjust itself in accordance with the level of stocks, making those who in fact have to hold this fixed stock willing to hold it at the level of price, high or low, in the market. Another example, not understood until Keynes analysed it, is the market for money—in the short period, the rate of interest must adjust itself until it makes people willing to hold the stock of money which is in fact available for them to hold. Yet another example is the Stock Exchange, in which so many amateurs have burnt their fingers. Its violent ups and downs relate to an almost unchanged quantity of stock, the amount of new issues in any one year being very small in relation to the quantity of shares already on the market.

Land has value because it can earn a rent. This is a basic economic proposition which many people still find it difficult to grasp. The individual may say, "I have to charge a rent because I paid a high price for this land." It is only on analysing the operation of the market as a whole that one finds oneself compelled to accept the conclusion that the causation works the other way. Land has no cost of production (apart from a few very exceptional cases), and sells at a price only because people expect to earn a rent from it (or want it for certain other reasons which will be discussed later). And the rents of different lands only originate because there are differences between them. The whole concept of rent implies that some land is practically valueless. The discovery of this basic principle was due to Ricardo, early in the nineteenth century. It should not be necessary to restate it in this manner, but some teachers of "land economics" still do not seem to understand the Ricardian principle.

The word "rent", unfortunately, has multiple meanings. In ordinary parlance, it means what the farmer actually pays to the landowner. In most cases, this represents payment for the use of buildings, roads, hedges, fences, drainage, and other fixed improvements, as well as for the land itself. At the same time it falls short of the economic definition of rent, as we shall see, in cases where land taxes and rates have been levied on the land. The economic definition of rent is a broad one, which includes the quasi-rents on certain factors other than land which, though not in absolutely fixed supply as land is, may be in a supply which cannot be substantially varied for quite a long period (such as houses, or services of skilled professional men). Rent is the amount by which proceeds actually received exceed the minimum amount which would have been necessary to evoke the supply of the factors of production required. In the case of agricultural land, the "minimum amount required to evoke its services" will be, in the long run, provision for maintenance, depreciation and interest on buildings and other fixed improvements. Anything else will constitute economic rent. If a land tax, rate, tithe, or similar charge be imposed on the land this must necessarily be regarded, in view of the above definition, as part of the economic rent of the land, though the economic rent receivable by the owner will be less to the extent of the tax.

Besides these differences of definition of economic rent and the rent of a farm as ordinarily understood, we know that the rent actually payable may be reduced to a level greatly below the economic rent by legislation of the sort which was enforced in Britain in the 1940s and

1950s forbidding landowners to raise rents or to terminate tenancies. When actual rents are kept below economic rents in this manner, there will, as an inevitable consequence, be a queue of unsatisfied applicants for farm tenancies.

A particularly glaring disregard of economic principles is shown by the British Ministry of Agriculture each year in its *Annual Review and Determination of Guaranteed Prices*, when it treats rent as a "cost" which has to be taken into consideration when determining the prices which a farmer should receive. It should be unmistakably clear that economic rents are in fact a *consequence* of the prices paid to farmers, less the payments incurred to evoke the required supplies of other factors of production (including the necessary remuneration of the farmer's own effort, risk-taking and skill). While rents remained at an arbitrarily low level, far removed from economic rents, this procedure at any rate did not do very much practical harm. But when actual rents approach economic rents, as they are now doing, the whole situation will become dangerously unstable. Every rise in prices, or economy of inputs, is likely to lead to a further rise in rents, which will then be used as a justification for a further rise in prices.

It is true that farmers' leaders in some other countries with high land rents have conducted their case in a similar manner. But one would have thought that the British Ministry of Agriculture, with all the economic *expertise* available to it, would not have shown such ignorance.

In Ricardo's view of the economy the rent of agricultural land played, by our standards, an unduly large part. This was understandable. At a time when many people were living not far from the hunger line, and there were no large imports of food in prospect, while population was increasing quite rapidly, the need to provide food posed the most important of all economic questions. Ricardo understood that food supply set limits to the possibilities of any increase in industrial employment. Recent work has shown that this is still precisely true for low-income countries at the present day.[1] In such countries, as agricultural productivity increases, a certain proportion of the increased product is used to increase standards of consumption all round. Some increase in product per head of the agricultural population is necessar-

[1] Clark and Haswell, *The Economics of Subsistence Agriculture*, chap. X.

ily required to feed any increase in the numbers of the non-agricultural population. The remainder is needed to provide exports to pay for urgently needed imports. It is found that an exponential increase in agricultural productivity is needed to provide for a linear increase in the proportion of the labour force engaged in non-agricultural pursuits—probably because, as the industrialisation of a country advances, the scales of food consumption rise progressively, as also do requirements of imports.

But whether we consider agricultural rents to be of major or of minor importance in the economy, Ricardo's original formulation requires supplementing in two ways, which are dealt with in subsequent chapters. One criticism was made in his own generation by Von Thünen (an agricultural estate manager, while Ricardo was a stockbroker), namely that Ricardo treated agricultural product as homogeneous. In fact, transport costs (which were very heavy in those days of horse transport) were the cause of substantial differentiations of agricultural production, and rents, in their turn, were the consequences of these differentiations. Geographers have paid altogether excessive attention to Von Thünen's principle, forgetting that its most striking applications were only valid in the days of horse transport. But economists, now as then, have paid it too little attention.

Ricardo's other deficiency was that (in common with all economists of his time) he did not think as clearly as we do in terms of marginal versus average costs, or returns to factors. These may substantially differ.

CHAPTER 2

VON THÜNEN'S "ISOLATED STATE"[1]

DURING the later years of the Napoleonic Wars (but apparently not disturbed by them), a practical, serious-minded, but remarkably imaginative German landowner, actively engaged in managing his estate, was reading Smith and Ricardo, disagreeing with them on a number of important points, and slowly preparing his own book on the theory of rents, transport costs, and land use. The writer was Von Thünen. He did not publish his book until 1827, by which time the simple Ricardian theory had obtained a firm hold in the intellectual world. In the later years of his life, which was long, he turned his attention to wage theory, developing some ideas which Sir John Hicks found of great interest. This part of his work is not now under consideration.

Professor Peter Hall recently made available a careful translation and annotation of his work, which most English-speaking readers have hitherto only known indirectly,[2] although he could have greatly increased its usefulness by translating the results into modern weights and measures (he gives the necessary conversion factors), and making some of the relevant comparisons with other agriculture, past and present. Von Thünen was far ahead of his time in the thoroughness and accuracy with which he kept records of farm and forest costs and proceeds, rents, and transport costs. Not only did he keep the records; he understood how to generalise the experience of his own and neighbouring estates, and to submit the results to economic analysis.

[1] The text of this chapter appeared originally as a contribution to *Oxford Economic Papers*, Nov. 1967. I am grateful to the Clarendon Press for permission to reprint it.

[2] Von Thünen's *The Isolated State*, edited by Peter Hall. Pergamon Press. £5.

5

Von Thünen's great achievement was to point out that transport costs were the cause, and rents the consequence, of important differentiations of agricultural, dairy, and forest production, according to distance from the market. Northern Germany in the early nineteenth century was an excellent testing-ground for such theories; it was a society which had advanced well beyond the subsistence agriculture stage, and was fully commercialised; while at the same time the high costs of transport in the pre-railway age called for sharp differentiations of production. Von Thünen was additionally fortunate in being able to prepare his studies in a region entirely dependent on land transport; he points out that his theorem would be upset in an area whose transport costs were reduced by the existence of a navigable river. The principal market town of his area, Rostock, was close to the sea. At that time some grain export from the region took place. But probably also there was some import of timber from Scandinavia, which would have affected his calculations.

Long-forgotten considerations are brought back to our minds. In those days, a large proportion of the total agricultural area had to be set aside to produce grain and fodder for the horses; the release of this land has been an important factor in raising the productivity of European and American agriculture during recent decades. At the same time, every town produced vast quantities of horse manure (not to mention the cow manure from town and suburban dairies, which survived until quite recent times) and all of this manure had to be disposed of without incurring excessive transport costs, which necessarily meant a small belt of market garden and milk-producing land in the immediate neighbourhood of the town of exceptionally high fertility, and also exceptionally high rent. Von Thünen quoted a contemporary British writer[3] as showing that horticultural land (including improvements) was being rented at £8–12 per acre per year on the outskirts of Edinburgh, and near London at a gross rent of £20 (of which rates and tithes took £10). The order of magnitude of average agricultural rents at that time was only about £1 per acre per year. Prices and rents in Germany were much lower, as we shall see; nevertheless, high-quality grazing for dairy cows was being leased on the outskirts of Hamburg for £10 per acre per year.

[3] Sinclair, *The Code of Agriculture*, 1817, p. 410.

Not only the cost, but also the delays of horse transport, made it obligatory that perishable products—milk and fresh vegetables—should be produced in the immediate neighbourhood of the town, on this very highly manured and very highly rented land. But Von Thünen also pointed out that wood, both for fuel and for building, could not stand high transport costs, and that it was therefore essential that some areas should be preserved for forests in fairly close proximity to each city. (The woodlands which still make the German landscape so attractive in the neighbourhood of cities were apparently not preserved solely for recreational reasons.) Owners of forests, however, Von Thünen concluded, must be running them as a public service rather than as a business; financially, they would be better off by clearing their land and investing the proceeds in fixed-interest securities, unless the rate of interest fell below 3 per cent. He appears to be thinking in terms of nearly a hundred-year growth cycle, for beech, also a favourite in neighbouring Denmark, a beautiful tree producing hard and valuable wood which we now use for furniture and similar purposes. It seems a mistake to use such a wood for ordinary building and fuel purposes; and perhaps already it was feeling competition in the market from Scandinavian softwoods. Much of the forest in Europe today is maintained by local authorities, and as a public service (for recreation, etc.) rather than as a business investment. At any probable rate of interest, the returns from forestry fall far short of the returns from agriculture on good land, indeed on any land that is not exceptionally steep or cold.

Apart from these limited areas required for the production of fresh vegetables and milk, timber and firewood, the city will be surrounded by an approximately circular zone of grain-growing land, right up to the limits permitted by its cost of transportation. At a time when a substantial part of the cost of transportation consisted of the feeding of the horses *en route*, and some of the men's wages were also paid in grain, Von Thünen measured transport costs in "natural units", a device which it has been found very useful to revive for analysing the problems of low-income farm economies.[4] The natural unit is measured by the actual quantity of grain fed to horses[5] on the journey and the

[4] Clark and Haswell, *The Economics of Subsistence Agriculture*, 2nd edition.

[5] A horse can live on grass and hay, but usually requires grain-feeding when he is working hard. In any case, hay would be too bulky and inconvenient a fodder to carry on a long journey.

return journey, together with the grain, or grain-equivalent of the cash expended on wages and other costs, per ton-kilometre of transport performed. Measured in this way, we can see directly how many kilometres a ton of grain can be transported before it has lost half (or any other specified proportion) of its value. In the case of products with a higher selling value per unit of weight than grain (e.g. butter and wool) the distance over which they can be transported rises in the ratio of their price to the price of grain; we expect these products therefore to be produced at a greater distance from the city.

The staple grain in North Germany at this time was rye, wheat being little used. All economic measures therefore are expressed in terms of kilograms of rye. (The nutritive value of rye bread is not significantly different from that of wheat bread; most people find the latter more palatable; many Germans on the other hand still prefer rye.) Rye was carried in large wagons, containing 1.16 tons when they reached their destination (more when they began their journey, to provide fodder on the way). On a representative journey of 37 kilometres length, costs worked out at 2.73 kg of rye-equivalent per ton-kilometre of delivered weight. Von Thünen calculated the absolute maximum distance from the town at which grain could be grown at 230 kilometres. At this distance, transport costs would reduce the proceeds to one-third of the town market price, which would leave no rent at all. Von Thünen then goes on to illustrate his principle by an interesting theoretical statement. "If potatoes were the only edible vegetable, arable farming would cease at 70 kilometres from the town; the Isolated State would be far smaller; and the town itself would have a smaller population."

Converting prices into British currency (one new thaler = 3.21 shillings) the price of rye at Rostock was 5.55 £/ton. Von Thünen's estimates for the estates in his neighbourhood, where the price of rye at farm was 4.78 £/ton, showed that the proportion of the gross proceeds (which included barley, oats, and pasture as well as rye) represented by rent (pre-tax) was 29 per cent, and a further 27 per cent represented a contribution to the farm overheads, including payment of interest on the buildings. This latter entry is unfortunately somewhat ambiguous. We have, however, some grounds for estimating the order of magnitude of the interest, depreciation, and maintenance costs on the buildings and equipment at 10 per cent, of the gross product, thus making rent, defined in the broadest manner, some 45 per cent of the gross product, i.e. cultivation costs, excluding all rent elements, as 55 per cent of price at Von Thünen's farm, or 47 per cent of the price in Rostock market. It appears at first sight that Von Thünen's figure of 230 kilometres as the distance limit set by transport costs for grain cultivation, which reduced the price at farm to one-third of the Rostock price, was too high. But this figure may be explained if we take into account (see p. 12 under dairying) the consideration that in the more distant zones labour may have received the same wage in terms of rye, but a lower wage in terms of money.

Evidence has been collected[6] from numerous sources, past and present, of transport costs in low-income agricultural countries, measured in the same natural units. The median cost of wagon transport is 3.4 kg grain-equivalent/ton-kilometre transported, making it cheaper than pack-animal transport (median cost 4.1) or human porterage (median cost 9.0), though much less advantageous than boat transport (median cost 1.0). Costs of wagon transport were found as low as 1.6 in the United States in 1800, where horses were exceptionally cheap, and as high as 8.0 for long journeys carrying army supplies in thirteenth-century England, where the driver may have had to take out the medieval equivalent of an insurance policy. Von Thünen's cost of 2.73 was obtained using his estate horses and labour. Beech logs were carried from the forest to the town by a contractor at a cost of 3.1, over a fairly short distance, not precisely specified. However, this was an exceptionally bulky load (7.8 cubic metres/ton) and possibly for grain transport the contractor's quotation might have been lower than the cost with estate horses. Butter, which had to travel a distance of 166 kilometres, incurred a transport charge as high as 7.5 per ton-kilometre. Presumably the carrier had to travel as fast as possible because of the perishability of his load; but in any case these long journeys always seemed to incur higher costs per ton-kilometre, probably because of the difficulty of persuading men to spend some time away from home.

We are given an interesting glimpse of the landscape beyond the 200-kilometre limit from a market town, whose inhabitants will grow a limited amount of grain for their own subsistence, but most of whose land will be given over to grazing, to produce butter and cheese which

[6] *The Economics of Subsistence Agriculture*, chap. 9.

can be sold, milk for their own consumption, and meat animals who can be walked to market. Still more distant zones were given over to the production of wool, whose value per unit of weight was some five times that of grain, and which would stand much greater transport costs. It was another achievement of Von Thünen's theoretical insight that he foresaw the development of wool production in Australia, as indeed took place, many years before the coming of railways.

Von Thünen gives us a most interesting collection of average prices of rye in German markets over the period 1828–41, showing "transport bias" whereby grain prices fall with distance from the market—and with them wages and other costs expressed in grain. Von Thünen himself took this consideration into account in calculating costs in the remote pastoral zone.

TABLE 1. *Rye Prices 1828–41, £/ton*

East and West Prussia	4.02
Posen	4.26
Silesia	4.55
Brandenburg and Pomerania	4.65
Saxony	5.07
Westphalia	5.84
Rhineland	6.30

Prices rose towards the western and more populous part of Germany, which was also nearer to the export market. Here, however, Von Thünen had a justifiable grievance. Britain, which had been a grain-exporting economy up to 1750, had become, with the rise both in population and in real incomes, a grain-importing country, subject to comparatively moderate tariffs, in the later eighteenth century. In 1815, the "Landlords' Parliament" made the Corn Laws more stringent.[7]

Germany was the principal source of exportable grain at this time. The effect of the Corn Laws was to create a great disparity in prices. During the period 1828–41, when the maximum German price of rye

[7] They had included, since the seventeenth century, with a view to stabilizing internal prices, the device now known as the "variable levy", a deplorable ancient error which the European Common Market authorities have revived, under the impression that they have hit upon something ingenious and new.

was only a little over £6 per ton, our ancestors were paying £14 per ton for wheat, and many of them were going hungry. Further analysis is required before we can judge the extent to which these adverse consequences fell upon German landowners and German labourers respectively; the relative movement of German wages and of German rents after the sudden repeal of the Corn Laws in 1846 should provide interesting material for study.

The next point to be noted is that on the estates which Von Thünen recorded productivity was exceptionally high by the standards of the time, indeed was at a level which Germany as a whole did not attain until a century later.

TABLE 2. *Crop Yields in tons/hectare*

	Wheat	Rye	Barley	Oats	Potato	Other roots	Hay
Mecklenburg 1810–40							
Von Thünen	2.14	1.98	(2.0)[h]	2.10	20.0	14.4	7.0[a]
France[b] 1781–90	1.15[c]	0.8	1.1	0.5	..	..	..
1815–24	0.82	0.65	0.84	0.73	..	..	..
England[d] about 1700	1.3	1.3	1.4	1.3	..	..	..
Germany, Bittermann[e]							
1800	1.03	0.90	0.81	..	8.0	..	..
1846–52	1.19	1.05	1.05	..	..	..	..
1878–82	1.46	1.16	1.58	1.32[g]	8.9	23.7[f]	..
1908–12	2.07	1.78	2.01	1.96[g]	13.3	28.0[f]	..

[a] Clover hay.
[b] Toutain, *Cahiers d'Isea* No. 115 (1961), p. 74.
[c] Spring wheat 0.9 and 0.8 respectively.
[d] Deane and Cole, *British Economic Growth*, p. 67. Richardson, *Outlook on Agriculture*, winter 1960, gives 1.4 for 1750, for which date, however, M. K. Bennett (*Economic History*, vol. iii) gives only 1.0, rising to 1.8 in 1850.
[e] *Die Landwirtschaftliche Produktion in Deutschland 1800–1950.*
[f] Sugar beet.
[g] *Agrarwirtschaft*, Apr. 1960 (for 1881–5 and 1910–14).
[h] Area not given, but approximately inferred from information on rotation.

Both the climate and the soil of Mecklenburg are favourable to agriculture; and from the book we can gauge the care and thoroughness devoted to manuring, rotation, and other methods then available for preserving soil fertility, in those days before chemical fertilisers. At that time, the chemical requirements for soil fertility were not even theoretically understood. Von Thünen made an heroic effort to compute fertiliser units, the requirements of different crops, and the amount of manuring required in consequence.

Until recently, Mecklenburg landowners had operated the old "three-field system". A representative estate covered 150 hectares —large even by modern standards—worked as a single unit, only one-eighth of the total area being sown to rye in any one year, which was expected to yield at the rate of 1.9 tons/hectare, or 36 tons of rye. The rest of the farm was in barley, fodder crops, pasture or fallow, yielding, after feeding its own livestock, further saleable products equivalent to another 40 tons of rye. Out of this gross yield equivalent to 76 tons of rye, he reckoned a pure rent (pre-tax) at the equivalent of 25 tons, and contribution to farm overheads, including paying 5 per cent interest on structures, at another 20 tons, which latter may have included, as stated above, some elements of economic rent. The remaining expenses included 7 tons of rye-equivalent for seed, 13 for cultivation, and 11 for harvesting—the two latter being predominantly payments for labour. At the beginning of the nineteenth century, progressive landowners were going over to the seven-year rotation, dividing their estates into seven equal areas for one year each of rye, barley, and oats cultivation, three years of pasture, and one for fallow. This raised the gross product on the same area to 111 tons rye equivalent. Rent per unit of area rose, but it now represented a smaller proportion of the gross product, at 32 tons; the contribution of overheads was again put at 27 per cent of the gross product. Von Thünen had received and scrutinised information from Belgium, the most densely populated and most productive part of Europe. If Belgian methods were applied in Mecklenburg, he concluded, they could raise the gross yield 2.2-fold and the rent 1.7-fold.

The figures for dairy production, however, tell a very different story. The average cow weighed only 250 kg, or about half the weight of a good modern dairy cow; and produced butter at the rate of only 42 kg/year, corresponding to a milk yield of only 900 kg (200 British gallons). In the Rostock market butter exchanged for eleven times its weight of rye—very similar to the ratio prevailing in the world market now. Even at the high transport cost already quoted (7.5 kg rye equivalent/ton-kilometre) butter could be carried 200 kilometres for less than 15 per cent of its selling price. A butter producer near the city would have incurred, quite apart from rent, direct production costs equivalent to 9.6 kg of rye in producing a kilogram of butter which he could exchange for 11 kg of rye. Of these costs, however, only 22 per cent were actually payable in cash (purchases of equipment, etc.) and the remainder were, in effect, payable in rye, namely payments for labour and contract services.[8] Farm workers further away from the city obtained exactly the same amount in rye for their labour service, but its money equivalent was less. Costs at a distance thus represented a smaller proportion of proceeds, leaving a larger amount for rent (in rye-terms: it would not be higher in money-terms). Or, if we like to put it another way, distant dairy-farming lands could earn rents which represented the fruit of the manœuvre of transporting rye values to Rostock in the form of butter, which was more compact to transport. There was, of course, a limit to this. The possibilities of producing butter and wool extended the effective boundaries of the Isolated State, but nevertheless in due course a limit must be reached at which rent fell to zero.

Von Thünen's tidy mind was offended, as indeed are those of agricultural economists throughout Europe to this day, by the ancient and irrational farm boundaries. He made some striking calculations to show precisely how much they increased farm costs. Rationalisation of farm boundaries, then as now, was hindered not only by the conservatism of farm families, and their attachment to their ancestral property, but also by a severe and vexatious system of taxation of land sales. Land should be taxed when people hold it without selling it, rather than the converse.

---

[8] Von Thünen clearly implies that wages were negotiated in rye-terms, not in butter-terms. Real wages, though above subsistence level, were low, and most of the butter output appears to have been consumed in the town. To contemporary readers, no doubt, these facts would have been too obvious to need stating explicitly.

CHAPTER 3

LAND VALUES FROM PRODUCTION FUNCTIONS

IT WAS in the year 1862 that a radically new epoch in economic thought began. This was the year in which Jevons[1] first made the distinction between average cost and marginal cost, now the foundation of economic teaching. It was Professor Peter Wiles who pointed out that the complete ineffectiveness of Marxian economic theory, its utter inability, in spite of the immense resources at its command, to analyse any concrete economic situation, arises simply from the fact that Marxian theory is pre-Jevonian, unable to distinguish marginal product from average product, marginal cost from average cost. Although Jevons was his contemporary, Marx apparently had never heard of him.

If we are not to be pre-Jevonian ourselves, therefore, we must replace the simple Ricardian theory by one which recognises that land has an average product and a marginal product which may differ, and that its rent should depend on its marginal product. It is possible for us to make a direct estimate of the marginal productivity of land when we have a production function, that is to say a mathematical expression for the expected product from varying inputs of land, labour, and other factors. We can obtain marginal productivities more precisely when we have mathematically prepared programmes for representative farms. Lacking either a production function or a programme, we have to fall back on estimating economic rent as a residual, from the gross product after all other necessary inputs have been remunerated. This method, in effect, assumes that the actual remuneration of the other inputs

---

[1] In a paper presented at the Cambridge Meeting of the British Association for the Advancement of Science (see Keynes's *Essays in Biography*).

has been according to their marginal value, which is not always the case.

These three approaches are dealt with in succeeding chapters, beginning with production functions. The number of good production functions for agriculture is still very limited. In any case, as we shall see, and indeed as common sense could have told us, the average and marginal productivities of land are very closely bound up with the extent of the inputs of other factors, and not only with the quantities of these other inputs (principally labour), but also with their price.

The classical Ricardian situation in which labour (and managerial enterprise) had to compete for a limited supply of land, with few or no opportunities of being employed elsewhere, and therefore the rent of agricultural land rose as a function of rural population density, is strikingly exemplified by the experience of several countries now, and by our own eighteenth-century past. In England this relationship between land rent and rural population density had completely disappeared by 1850. Nothing like it is to be expected in modern industrial countries, in which labour and enterprise can find employment in industry if they do not like the terms which agriculture offers.

The following diagrams (Figs. 1–5) show clearly the relation between agricultural rents, or land prices, and rural population density in Italy (Fig. 1),[2] pre-Communist Romania (Fig. 2),[3] the Philippines (Fig. 3),[4] Poland (Fig. 4)[5] and eighteenth-century England (Fig. 5)[6]—all countries in which, on the whole, agricultural labour "had nowhere else to go"—though the situation in Italy is now rapidly changing, and land prices are falling fast. It will be seen that in Lazio (the country surrounding Rome) and Marche (on the Adriatic coast where the gas-

[2] *Annuario dell'Agricoltura Italiana*, 1960, pp. 144–6.

[3] Warriner, *Economics of Peasant Farming*.

[4] Land price data from National Bank of the Philippines, employment from Census.

[5] *Statistical Yearbook* (in Polish). I am indebted to Mr. H. Frankel for translation and interpretation.

[6] Rents from Caird, *English Agriculture, 1850–51*, p. 474, quoting Arthur Young, converted to wheat equivalents. Area of agricultural land assumed to be the same as that shown by the first agricultural returns in 1867, excluding rough grazings. Agricultural population from Deane and Cole, *British Economic Growth*, p. 103. Male agricultural population assumed at 25 per cent, 20 per cent and 17½ per cent of total in counties classified as agricultural, mixed, and industrial respectively.

field has been discovered) land values are much below expected, probably because so much alternative employment is available.

Although Poland is a communist country, land can be freely bought and sold by owner-occupiers (as it can also in Yugoslavia); and indeed the price of land in real terms is now higher than it was in the 1930s. Unlike the position in Italy, proximity of industrial centres does not appear to lower the price of land, at a given rural density, and indeed proximity to the mining area of Katowice appears to raise it.

The objection has been raised that land values are not determined solely by agricultural population density but that the values are the result of climate, soil and relief in a particular area. By this argument,

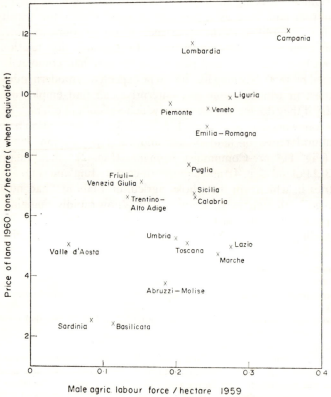

FIG. 1. Land values in Italy, 1960.

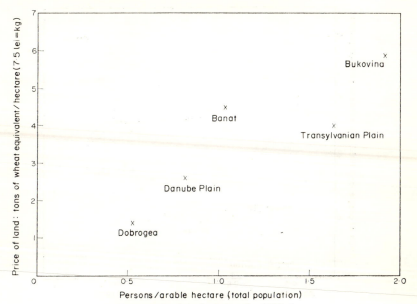

FIG. 2. Land values in Romania, 1923–7.

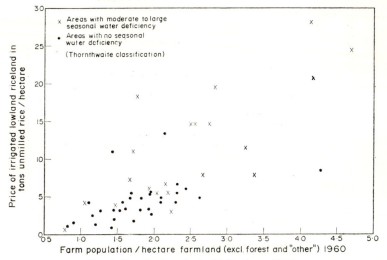

FIG. 3. Land values in the Philippines, 1967.

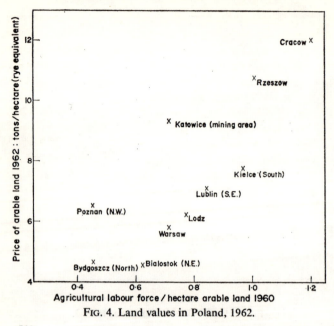

FIG. 4. Land values in Poland, 1962.

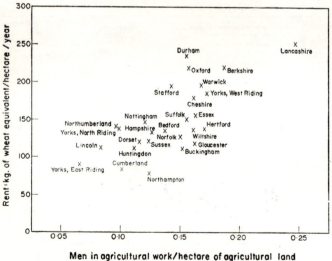

Men in agricultural work/hectare of agricultural land

FIG. 5. Rents and agricultural population density, England, 1770.

it is the environmental factors which account for the land value, and the population density is their consequence. This objection does not appear to be valid if the diagrams (Fig. 5) such as that for eighteenth-century England are examined. A more thorough test has been made by examining the relationship between land values, population density and climate in the irrigated land of the Philippines. The fifty-six provinces were divided into two climatic zones on the basis of Thornthwaite's classification, showing the regions of moderate or large winter water deficiency, and the regions of small or no seasonal water deficiency. The resulting diagram shows that there was a definite relationship between price of irrigated land and the population per hectare, irrespective of climatic zone.

The Romanian diagram (Fig. 2) shows a slope of approximately 3, i.e. a rise in land value of 3 tons/hectare for each additional person/ hectare. It is very interesting that an exactly similar coefficient should have been estimated in 1897 for Chile,[7] namely that a density increase of 1 person/km^2, i.e. 0.01 person/hectare, raised land value by 0.03 ton wheat equivalent/hectare (i.e. 3 pesos/ha.).

Another aspect of this situation is shown in the writings of Leroy-Beaulieu, a French economist of the 1880s whose work received little attention from his contemporaries—perhaps because he wrote in language understandable by ordinary readers. Challenging the crude Ricardianism of that time, which tended to assume that rent would represent the highest proportion of product on the best land, he showed that the opposite was in fact the case. Rent took the lowest proportion of product in Département Nord, where the land was good but there was also abundant alternative industrial employment, and most in Cantal and Aveyron, poor mountain regions where there was no alternative employment.

Work on agricultural production functions has developed during recent years. The Cobb–Douglas function, criticised though it has been by theoretical econometricians, has proved surprisingly durable, and is now generally used. (I knew both Cobb and Douglas personally in the 1920s.)

Though most readers are familiar with the Cobb–Douglas function, it may be necessary to explain it briefly to those who are not. It was

---

[7] Kärger, *Landwirtschaft und Colonisation in Spanischen Amerika*.

originally designed for explaining the time-series of the quantum of industrial production in Massachusetts, in terms of varying inputs of labour and capital. It was found that the quantum of labour input to the power a, multiplied by the quantum of capital input to the power b, multiplied by a constant, gave a good explanation of quantum of product. The exponents a and b were both between 0 and 1, and their sum came close to 1. It can be shown mathematically that if economies of scale prevail, i.e. if an all-round increase in inputs, maintaining their previous proportions among themselves, leads to an increase in output in *greater* proportion than the increase in inputs, then the sum of the exponents will exceed 1; under diseconomies of scale, it will fall short of 1; a figure of 1 indicates neither economies nor diseconomies. It is generally held (though not universally; particularly there are exceptions among some of the most sophisticated industrial economists) that industrial production is subject to economies of scale—though these may take some years to show themselves. In the case of agriculture, we are all thoroughly used to the idea of an increase in the input of one factor, other inputs remaining given, yielding diminishing returns. When we try to imagine a situation (which rarely if ever occurs in fact) of an agriculturist increasing all inputs simultaneously, in uniform proportion, few, if any of us, expect "economies of scale", i.e. an increase in output in greater proportion than the increase in inputs—except perhaps in a few anomalous cases. In constructing production functions for agriculture, therefore, the mathematical constraint that the exponents should add up to 1 is frequently though by no means universally introduced.

While the Cobb–Douglas function was originally designed to explain output in terms of only two inputs it has the great merit of being capable of extension without difficulty to cover simultaneously as many inputs as the data, and computational facilities, will stand. For operation the function is conveniently transformed into the logarithm of the product being expressed as the sum of the logarithms of the inputs, each multiplied by an appropriate coefficient, plus a constant. Partial differentiation enables us to estimate the marginal productivity of each input, customarily expressed when the input is at its mean (arithmetic or geometric) value; but if we wish, we can draw diagrams showing marginal productivity over any desired range of values. If each factor is rewarded according to its marginal productivity, in the absence of economies of scale, the exponent for the factor will indicate the share of the total product which it will receive.[8]

There has been a most fruitful development during recent years, in the field in which Professor Heady is the undoubted leader, of preparing production functions for agriculture in economically advanced countries, sometimes from time-series, but more usually from cross-section analysis of a substantial number of farms for which full records are available, with the number of inputs analysed sometimes rising as high as six. Inputs of equipment, fertilisers and feeding stuffs are often deemed worthy of analysis, as well as of capital, labour and land; one feels the regret on the part of econometricians that they cannot yet find a way of quantifying enterprise and managerial skill to go into the computer too.

When, however, we come to production functions for low-income countries, the input of fertilisers has been (at any rate until recently) inconsiderable, likewise the input of equipment and purchased feeding stuffs. However, their type of agriculture demands production functions in terms of land, labour, bullock power (of critical importance in India and Pakistan in view of the shortness of the monsoon season) and, in many areas, of the quantity of irrigation water.

It is when we turn to the simplest type of agriculture, which still prevails over nearly all of Africa, and some of Asia and Latin America, namely hand-hoe cultivation, where expenditure on equipment is insignificant, and on fertilisers and feeding stuffs zero, that we can construct the simplest type of agricultural production function, explaining output in terms of inputs of land and labour only. Economies of scale are clearly absent, and the marginal products of the two factors, multiplied by the quantities of inputs, must add up to the total product. Furthermore, agriculture in these areas is virtually the sole source of livelihood, and there is no significant competing demand for either labour or land. Let us examine this simple situation first before we proceed to the more complex ones.

Study of production functions for this type of agriculture began in the early 1950s with fitted algebraical functions for cocoa and for food

---

[8] Douglas told me that when the function was first prepared in the 1920s, he was expecting it to show that the wages then actually received by labour were considerably below its true marginal product; and was surprised to find that they were in fact extremely close to the level predicted by the function.

crops in Nigeria,[9] and empirically tabulated results for groundnuts and other crops in Gambia.[10] The Nigerian equations were

Cocoa yield (lb) $= 53.5 \text{ acres}^{0.67} \text{ Man-hours}^{0.33}$
Food crop yield* (shillings) $= 2.68 \text{ acres}^{0.24} \text{ Man-hours}^{0.76}$

* Net of seed requirements.

The exponents are seen to differ widely. However, in the case of cocoa, the "acreage" variable also includes, in effect, a large capital input—the cocoa bush takes a long time to come into bearing. The food-crop equation indicates that, if the factors are remunerated precisely according to their marginal productivities, labour will receive 76 per cent of the product net of seed requirements, and land 24 per cent.

This latter appears to indicate the general African order of magnitude. Haswell's results for growing groundnuts in Gambia indicated marginal returns to labour actually rising at the lower levels of input, to a maximum at an optimum labour input of approximately 570 man-hours/hectare/year, but falling very rapidly as the labour input approached 1000 man-hours/hectare/year.

Marginal and average returns to land were:

Land supply, hectares/000 man-hours	0.94	1.22	1.75	3.11
Average product groundnuts,[11]				
kg/000 man-hours		755	855 855	870
Marginal product, kg/hectare			358 0	11

On the smallest holdings the marginal product of land is about 45 per cent of the average product, but on the larger holdings it disappears.

A considerable step forward in the analysis of African agriculture was made by Massell and Johnson[12] when they pointed out that, under African technology, labour requirements for cultivation were in effect precisely determined by the amount of land cultivated, labour

[9] Galletti, Baldwin and Dina, *Nigerian Cocoa Farmers*, 1955, pp. 314 and 328.
[10] Haswell, *Economics of Agriculture in a Savannah Village*, Colonial Office Research Study No. 8, 1953.
[11] Undecorticated.
[12] Rand Corporation Publication R/443/RC.

requirements for harvesting by the weight of the crop, and that the only flexible element in labour input was that devoted to weeding. This they analysed separately. They studied two areas in Rhodesia, the Chiweshe area of traditional tribal agriculture, and the Darwin area, where a small number of selected and trained African farmers were given much larger holdings which they held freehold, and not subject to tribal tenure. In the tribal area the marginal product of a hectare of groundnuts was only 62 kg, of millet 112 kg, or some 10–20 per cent only of average products. However (judging by the results from Gambia), these averages may conceal a high marginal productivity on the small holdings and a very low productivity on the larger.

Maize showed a much higher marginal productivity per hectare, for the paradoxical reason that this was an export crop facing very high transport costs. Its internal price in Rhodesia was therefore less than a third of the price of millet, whereas on world markets maize is generally priced higher than millet. As a result of this situation, maize in Rhodesia is only grown on the best soils.

When we examine the returns in the modern farming area, we get the striking result that the marginal productivity of *land* is raised five-fold for millet, seven-fold for maize and ten-fold for groundnuts, in comparison with the tribal area. The marginal productivities of labour, capital and fertilisers rise in much smaller proportions (Table 3).

Further production functions for areas of simple hand-tool agriculture have recently been compiled by Luning.[13] For a densely populated area in Northern Nigeria, with sandy soil and inadequate rainfall, growing groundnuts and subsistence crops, with a labour input averaging some 400 man-hours/hectare/year (i.e. lower than the optimum labour input found by Haswell in Gambia), he presents an equation for product/land as a function of man-hours/land, which shows a very small marginal productivity of land. The principal factors affecting yield are the purchased inputs other than labour and land, i.e. seed, fertilisers, and insecticide. Here again, however, this result may be deceptive—these inputs will be strongly correlated with the skill, education, and enterprise of the farmer.

---

[13] *Economic Aspects of Low Labour-income Farming*, Wageningen Agricultural Research Report No. 699.

TABLE 3. *Marginal Productivities in Rhodesia*
(Massell and Johnson)

	Tribal tenure area (Chiweshe)			Freehold area (Darwin)		
	Maize	Gro- und- nuts	Millet	Maize	Gro- und- nuts	Millet
Returns measured in kg crop/ physical unit input:						
Land (hectare)	254	62	112	1800	581	580
Weeding labour (hour)	0.41	0.26	0.38	0.95	0.33	1.42
Farmyard manure (ton)	108			48		
Returns measured in kg wheat equivalent/physical unit input						
Land (hectare)	190	68	76	1350	639	394
Weeding labour (hour)	0.31	0.29	0.26	0.71	0.36	0.97
Farmyard manure (ton)	81			36		
Returns measured per unit of money input						
Fixed capital (net)	0	0.116	0.033	0.132	0	0.083
Chemical fertilisers	1.69			2.94		
Average price of crops £/ton	10.5	42.5	33.6	10.5	42.5	33.6

Among rice-growers in Surinam in South America, on the other hand, with a similar labour input, the response to these latter inputs was inappreciable, while the marginal productivity of land appeared to be high. The equations were:

(i) Nigeria:

$$\log(\text{shillings product/acre}) = 1.92 + 0.017 \log (\text{acres}) + 0.340$$
$$\log (\text{man-days/acre}) + 0.829 \log$$
$$(\text{shillings purchased inputs/acre}).$$

(ii) Surinam (mean of separate determinations for two years 1965 and 1966):

bags paddy produced $= 2.32$ hectares$^{0.65}$ man-days$^{0.39}$ (purchase of seed fertilisers and insecticides, in guilders)$^{0.01}$.

The Nigerian equation gives a sum of the exponents of 1.186, indicating returns to scale, though statistical testing does not rule out the null hypothesis. This equation predicts a marginal product of labour of 1.6 shillings/man-day, as against an observed wage of 2.2. The Surinam equation predicts a (very much higher)[14] marginal product of 3.1 guilders/day, as against an observed wage of 3.3. The standard errors, however, are fairly high. The very high marginal productivity of land predicted by the equation, namely 200–300 Surinam guilders/hectare/year, is also confirmed in fact.[15]

It is surprising indeed to find that there has recently been some discussion[16] on the subject of marginal returns to labour in rice-growing in Communist China.

A case was quoted from Chekiang as seen in Table 4. The marginal return was computed by fitting a quadratic function.

TABLE 4

Labour input man-days/ hectare	Product kg paddy/hectare	Marginal return kg paddy/man- day
240	3000	28
330	5272	15
450	5625	0

If we can accept these figures, and reverse them to compute the marginal product of land, i.e. assume a given labour input of 450 man-

---

[14] The Surinam guilder is worth 0.53 $U.S. (not the same as the Netherlands guilder) and will purchase 11.8 kg of rice, unmilled weight.

[15] Luning, private communication.

[16] Ishikawa, *Economic Development in Asian Perspective*, pp. 257–8.

days cultivating 1, 1.364 and 1.875 hectares respectively, we obtain at the first stage a marginal product of 4.3 tons/hectare, or about three-quarters of the average product. But further land input shows a negative marginal product, i.e. the available labour is insufficient (with Chinese methods) for adequate cultivation.

To summarise, we must leave hand-hoe agriculture in a state of uncertainty about the marginal productivity of land, but with some indication that it may change suddenly, as observed in Gambia, once a certain minimum average size of holding is passed.

Japan is not precisely in this category, although until recently, at any rate, most of the agricultural work was performed with hand tools. Tutiya[17] prepared a production function for the Province of Shizuoka in 1951 (apparently imposing the restraint that the exponents should add to 1), and obtained exponents of 0.56 for land, 0.25 for capital and 0.19 for labour.

The much higher exponent for land is remarkable. A smaller study[18] in Kagoshima, the poorest Japanese province, indicated an exponent for land as high as 0.72.

The ratio of labour's marginal product to its average product in Japanese agriculture, which before 1940 averaged 0.34, has averaged 0.71 since 1950.[19] It should follow that the corresponding ratio for land should have fallen.

We now turn to Indian-type cultivation, where draft animals are used. First we have an equation for Ceylon, where land is comparatively abundant, which shows a low marginal productivity:[20]

$$\log \text{product (bushels paddy)} = \log (\text{acres}) + 0.577 \log (\text{man-days}) + 0.245 \log (\text{rupees expenditure on other inputs}) - 0.014.$$

Here the sum of the exponents adds up to substantially less than 1, indicating diminishing returns. The marginal product of labour indicated by this equation is substantially below the wage actually prevailing.

---

[17] *Quarterly Journal of Agricultural Economics*, 1955, No. 1 (in Japanese).
[18] Maruta, *Memoirs of the Faculty of Agriculture*, Kagoshima University, 1956.
[19] Minami, *Hitotsubashi Journal of Economics*, June 1970.
[20] Sarkar, *Journal of the Royal Statistical Society*, Part 2, 1957.

In India, a variety of results for the marginal product of land have
been ascertained.

TABLE 5A

	Marginal product of land rupees/ha./yr
Allahabad, 1955–6	394
Gujarat, irrigated rice, 1966–7	495
Orissa: 5-ha. farm	63
7-ha. farm	5
Uttar Pradesh, 1963–4	175
Near Delhi, 1961–2	460
Baroda, 1960 1	247
Ganges Valley	108

Sources:

Allahabad   Wycliffe, *Indian Journal of Agricultural Economics*, Jan.–Mar. 1959. Forty-eight farms, geometric average area 1.47 ha., employment 34 man-months, equipment (including draft animals) 308 R, other livestock 377 R.

Gujarat   Adhvaru and Parikh, *Studies in the Economics of Farm Management*, Surat and Bulsar (Sardar Patel University).

Orissa   Das Gupta, *Indian Journal of Agricultural Economics*, Oct.–Dec. 1961. Rice land.

Uttar Pradesh S. P. Dhondyal, D.Sc. Thesis, University of Delhi, 200 farms.

Near Delhi   Singh and Hrabovsky, *Indian Journal of Agricultural Economics*, Oct.–Dec. 1965.

Baroda   Naik, *Indian Journal of Agricultural Economics*, July–Sept. 1965.

Ganges Valley Hopper, *Journal of Farm Economics*, Aug. 1965.

The value of 460 rupees/hectare/year, or nearly a ton of wheat, was about the *average* yield for India as a whole, but this was presumably on irrigated land. The figure for Gujarat was for a time of very high prices, when unmilled rice sold for 950 R/ton at farm.

In the careful study which Hopper prepared for one village in the Ganges Valley, where water was treated as a separate input, the marginal product of land was only 108 rupees/hectare/year—quite a small proportion of average gross product. The marginal product of water was 0.25 kg wheat/m³ water, a figure of basic importance.[21]

Dhondyal's study in Uttar Pradesh referred also to land which was partly irrigated. The water was presumably applied to the crops speci- fied below, as they showed a much higher marginal productivity than the farm average (unfortunately we do not know the aggregate depth of irrigations during the season; if it were 25 cm, this amount of water would contribute, using Hopper's coefficient, 625 kg/ha. wheat to the marginal product, apart from the land). In this study, the summation of the exponents indicated diminishing returns to scale

TABLE 5B

Crop	Marginal product	
	kg/hectare	rupees/hectare
Wheat	812	406
Wheat and gram (pigeon pea)	842	438
Jowar and arhar (sorghum)	1050	535
Paddy	1230	449

Krishna[22] adopted the convention of treating irrigated land as worth 5/3 times the same area of unirrigated land and obtained for 1954–7 a marginal value for unirrigated land of 108 rupees/hectare/year —exactly the same as Hopper. At current prices this represented 0.28 ton wheat equivalent/hectare/year. Krishna found no significant returns to scale.

The low figures for Orissa are for a remote, backward and compara- tively sparsely populated state, in which there is some unused land.

Shingarey and Waghmare[23] prepared a production function for a high-yielding new variety (Taiching Native I) of rice, which on small farms (average 0.25 hectare rice land) showed a marginal return to land of unmilled paddy of 1458 kg/ha/year, far higher than any of the results quoted above. The marginal return to labour, however, was

[21] See Clark, *The Economics of Irrigation*, Pergamon Press.
[22] *Indian Journal of Agricultural Economics*, July–Sept. 1964.
[23] *Indian Journal of Agricultural Economics*, Oct. 1968, p. 64.

only 4.3 kg/man-day. Wellisz and others,[24] by regressing data for twenty districts of Andhra Pradesh over a period of 9 years, and introducing soil type as a dummy variable, found (at 1955–6 prices) a marginal productivity for unirrigated land of only 72 rupees/hectare/year, but 400–500 for canal-irrigated and 570 for tank-irrigated land. The logarithmic function gave the marginal productivity of labour at 497 rupees/man-year as against a wage of 305. It appears (though it is not certain) that increasing returns to scale prevail.

Before we leave India we may refer to an interesting study by Sen[25] comparing the Indian States, where density of farm population/hectare of "standard" farm land ranges from 1 in Assam to $2\frac{1}{2}$ in Bihar. By fitting a double-log relation between this variable and production/man-year, Sen obtains, in effect, a Cobb–Douglas function, showing

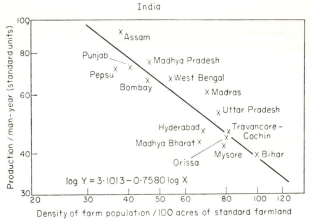

FIG. 6. Production/man-year (standard units).

an exponent for labour of only 0.24, of land 0.76. Definitions of "standard" farm land and "standard" units of product were not given.

The fitting of Cobb–Douglas functions calls for data from individual farms, or, at any rate, aggregates from widely different regions, as in

[24] *Journal of Political Economy*, July–Aug. 1970.
[25] International Conference of Agricultural Economists, 1952.

Sen's results just quoted. Less satisfactory, though still useful if better data are not available, are aggregates from farms grouped in size-classes (this treatment inevitably "wastes" some of the information). A very interesting result in this form was that obtained by Bićanić for Jugoslavia.[26] These results are particularly interesting because they refer to some of the most densely populated agricultural land in

TABLE 6. *Jugoslav Farm Families*

Size range of holding, hectares	Under 2	2–3	3–5	5–8	Over 8
Average size of holding, hectares	1.4	2.5	4.0	6.5	11.7
Labour supply man-year equivalents[a]	2.68	3.02	3.40	3.53	3.87
Labour supply excluding non-farm employment	2.34	2.74	3.17	3.31	3.71
Net farm product tons wheat equivalent/man-year	1.49	1.76	2.06	2.44	2.97
Hectares/man-year	0.60	0.91	1.26	1.97	3.15
Marginal product tons wheat equivalent/hectare		0.87	0.86	0.54	0.45

[a] Women's equivalent 0.7.

Europe. On the smallest holdings the net average product is 2.5 tons wheat equivalent/hectare and the marginal product about a third of this.

For another densely populated country in Eastern Europe we had a pioneering study by Pytkowski.[27] The marginal return on land in Poland was found to be the equivalent of only about 0.15 ton wheat equivalent/hectare, alike on the smaller and the larger holdings, and the author's primary concern was to try to persuade farmers to invest their savings in more livestock rather than in more land.

We now turn to the situation in Western Europe. Estacio[28] made

---

[26] International Association for Research in Income and Wealth, 1957 Conference.

[27] Warsaw Biometric and Statistical Laboratories, Vol. 2, 1932.

[28] *International Explorations of Agricultural Economics.*

inter-provincial comparisons for Portugal which showed a marginal productivity of "standard land" (actual data were weighted for vineyards and other specially productive lands) of only 43 kg wheat equivalent/hectare/year (87 escudos, or $3). Here land seems to be abundant.

We have a very thorough study of marginal productivities of different types of land in Belgium by Bublot.[29] In 1961-2 average factor-income/hectare of arable and pasture land was 21,000 francs or 4.5 tons wheat equivalent—the *gross* yield from land devoted to wheat was a little higher. Marginal product ranged from 24,000 francs/hectare on the best soil, *limoneuse*, to 11,000 on the worst, *condroz*; but the median appears to indicate a marginal productivity of some 80 per cent of average productivity. Under these circumstances, the marginal productivity of labour is bound to be low, though the wages paid are in fact considerably above marginal productivity on all soils except the polder. But Belgian agriculture mostly used family labour.

Some interesting results were obtained for English dairying in a detailed production function obtained for a number of farms by Imperial Chemical Industries.[30] It was admitted, however, that the farms were unrepresentative in having received exceptional amounts of fertiliser, and also unusually good management. An approximate analysis of the nationally more representative Farm Management Survey showed that, in the farm survey by I.C.I., marginal returns to fertiliser were some 2.2 times, to labour and purchased feeding stuffs some 1.5 times the national average.

That such high marginal returns to fertiliser could be obtained when using quantities so much above national average is indeed remarkable, and shows the importance of good management.

The functions were computed per acre of land, which, in effect, amounts to assuming in advance an homogenous function, without advantages or disadvantages of scale within the range of farm sizes studied. The equation below has been restated after multiplying throughout by acreage, which then gives exponents necessarily adding up to 1. Averaging the separately computed functions for the years 1953 and 1954, which were not appreciably different, we obtain the following:

[29] *L'Exploitation agricole*, p. 559.
[30] *Profits from Dairy Farming*, 1956.

Gross product from grazing stock, £/year $= 5.1$
Purchased fodder, £/year$^{0.24}$
Input of pure nitrogen, lb/year$^{0.19}$
Labour and machinery input, £/year$^{0.35}$
Acres of land$^{0.22}$.

Thus the marginal product of the land should be 22 per cent of the gross product.

This formula leads to some interesting conclusions. With land given, the marginal return to nitrogen input *rises* with the fodder input—presumably because the latter enables the farmer to keep more cattle on the land, and thus eat off the grass more efficiently. If the nitrogen input is fixed at 40 lb/acre the optimum fodder input is 11 £/acre. For nitrogen inputs of 60 lb and 80 lb the optimum fodder inputs rise to £13 and £15 respectively. In the latter case the marginal product works out at 47 £/ton of ammonium sulphate (approximately 20 per cent nitrogen) when its price was only £18.

The reason was that inputs of labour and machinery combined exceeding 17 £/acre/year (at prices of 1953/4) would be uneconomic. At that time the average earnings of a man were about £400 and the annual charges on the amount of machinery associated with him about £170. In other words, any land input less than 570/17.5 or 33 acres/man was uneconomically low.

Marginal products in Sweden have been thoroughly analysed by Sandqvist[31] for different regions and sizes of farms.

The marginal productivities are expressed in physical terms, in wheat equivalent tons/hectare, as Swedish prices are substantially higher than in most other countries. The results relate to 1963, for which year the price of wheat was taken at 450 kronor/ton.

Marginal returns to land are strikingly higher in the better soil and warmer climate of the South. These differences, however, are not fully reflected in the land values, which are approximately 6400 kronor/hectare in the south, and 2700 in the other two regions. Sandqvist's results, probably the most comprehensive series of production functions ever prepared, do, however, help us to analyse the interesting situation which economic theory would predict, namely that economically

---

[31] *Acta Agricultura Scandinavica*, 1963.

excessive inputs of other factors (indicated by low marginal returns to them) might put farmers in a situation of being willing, at any rate temporarily, to bid for land at a price higher than that indicated by its true marginal productivity, in order to obtain better employment of underemployed labour, equipment, etc.

In most cases marginal returns fall as size of farm is increased. This indicates that the smaller farms, for whatever reason, and whether temporarily or permanently, have inadequate inputs of land, or of the other factors in question. This is true of fertiliser inputs among small farms in the South and Forest districts. But in Central Sweden fertiliser inputs appear to be excessive. Crop labour appears to be employed to the economically optimum extent on the larger farms in the South, but to be seriously in excess in Central Sweden. Labour in livestock production, on the other hand, yields almost no marginal return in two regions, and a negative return in the third (table 3). The latter may be interpreted as an indication that the employment of labour for looking after livestock positively interferes with the performance of other potentially more remunerative activities.

The input of purchased fodder seems to be inadequate on the larger farms and in the forest districts. The price of oats in Sweden in 1963 was 0.4 krona/kg, suggesting that it was worth while producing more home-grown fodder in the forest regions, but not elsewhere, and that in Central Sweden such production actually impeded more remunerative operations.

The Central region clearly should have more livestock. In the other regions the returns were about in equilibrium, bearing in mind that they have to cover depreciation as well as interest.

Sweden appears to be seriously over-supplied with equipment, which is perhaps not surprising. The supply of buildings appears to be inadequate in the South, excessive in the other regions.

We have here also what appears to be the first attempt to measure the effect of the age of the farmer upon the productivity of the farm. This is comparatively uniform between regions, as it should be, and also, expectedly, is higher on the larger farms, though it does not rise in proportion to the size of the farms. On the large farm, each additional year for which the old farmer stays on is equivalent to the loss of half a hectare of land in the South, or 2 hectares of land in the forest districts.

TABLE 7. *Marginal*
Computed at geometric

Region: Farm size, hectares:	Southern plains, Gotaland				
	5–10	10–20	20–30	30–50	50–100
Marginal product of land, tons/ hectare wheat equivalent	1.98	1.61	1.48	1.27	1.20
Do. kronor/hectare	893	725	669	570	538
Marginal returns to other inputs (money return/money input)					
Fertiliser	1.73	1.51	1.31	1.24	0.99
Labour (crops)	0.85	0.94	1.03	0.96	1.02
Labour (livestock)	0.02	0.02	0.03	0.03	0.05
Purchased fodder	1.08	1.41	1.68	2.00	3.03
Equipment (annual cost)	0.30	0.30	0.28	0.28	0.26
Buildings (annual cost)	1.39	1.44	1.51	1.84	1.68
Miscellaneous inputs	1.86	1.74	1.66	1.60	1.45
Annual marginal return/capital value					
Livestock	0.20	0.20	0.23	0.26	0.39
Other marginal returns:					
Home-grown fodder, kronor/kg	0.22	0.23	0.24	0.22	0.20
Forest land, kronor/hectare					
Effect of farmer's age, kronor/year	− 63	− 92	− 150	− 185	− 249

Finally, it is interesting to see production functions for Australia, where land is abundant. For dairy land in northern New South Wales in 1952–3 (average size of farm 77 hectares), Dillon[32] found exponents (without imposing the constraint that they must add up to 1) of 0.28 for land, 0.22 for labour, and 0.42 for operating capital. The sum of the exponents was 0.92, indicating diminishing returns. For drier sheep country further inland, where the average holding was over 600 hectares, the exponent for land fell to 0.10, for labour rose to 0.59, and for capital to 0.55, giving a total of 1.24, indicating strongly increasing returns. Those engaged in buying and selling land, however, appear to have taken these facts and other prospects into account, making the price below marginal product in one case, and much above it in the other.

[32] Economic Society of New South Wales, Monograph 188 (1956).

Productivities in Sweden
mean of input values

Central Plains, Svealand					Forest Districts				
5–10	10–20	20–30	30–50	50–100	5–10	10–20	20–30	30–50	50–100
0.84	0.59	0.51	0.42	0.45	0.65	0.49	0.44	0.38	0.27
378	264	230	191	203	292	222	197	172	120
1.06	0.80	0.62	0.65	0.69	2.02	1.52	1.20	1.04	0.88
0.32	0.50	0.48	0.55	0.62	0.77	0.85	0.87	1.13	0.67
−0.44	−0.62	−1.03	−0.94	−1.09	0.04	0.05	0.06	0.09	0.07
0.86	0.91	2.27	1.75	1.22	1.61	1.79	1.65	1.54	1.53
0.74	0.63	0.59	0.59	0.60	0.33	0.34	0.31	0.31	0.30
0.20	0.3	0.27	0.28	0.38	0.60	0.63	0.66	0.71	0.72
0.49	0.37	0.36	0.32	0.31	0.66	0.75	0.63	0.57	0.53
0.83	0.81	1.04	0.89	0.66	0.26	0.24	0.24	0.27	0.24
−0.01	0.07	−0.21	0	−0.19	0.67	0.55	0.53	0.49	0.54
					83	106	130	153	210
−66	−142	−160	−202	−262	−51	−76	−109	−156	−236

Conversely, labour had a marginal product below its wage in dairying, and a marginal product above its (higher) wage on sheep properties. In both cases there appears to have been an over-optimal supply of working capital.

Duloy[33] made an analysis of production functions for wool-producing properties (a very small proportion of the product was contributed by livestock sales) in high- and low-rainfall areas in New South Wales in 1954–5. For convenience product was measured net of shearing and selling expenses. He calculated optimum inputs, on the assumption that the average grazier was free to spend only an amount equal to the total of inputs shown in the first column, but could rearrange as he wished between inputs (in fact his total inputs were higher, because buildings and machinery had been excluded from the analysis).

[33] *Australian Journal of Agricultural Economics*, Dec. 1961.

TABLE 8

Number of farms:	Richmond–Tweed area, Dairying, 1952–3				South-western Slopes area, Sheep, 1953–4			
	99				52			
	Exponents	Average input	Marginal product at mean of inputs	Price	Exponents	Average input	Marginal product at mean of inputs	Price
Land (acres)	0.28	191	£3.08	£2.5 (annual value)	0.10	1516	£0.39	£1 (annual value)
Labour (man-months)	0.22	21.7	£22.8	£46.5	0.59	23.1	£130.7	£60
Operating capital, £	0.42	702	1.47[a]	1.05[b]	0.55	1457	2.07[a]	1.05[b]
Sum of exponents	0.92				1.24			
Product, £		2300				5908		

[a] Ratio to input.   [b] Assumed interest rate.

The sum of the exponents for the low-rainfall zone was 1.26, indicating strongly increasing returns to scale, or, to put it another way, that the average enterprise was undersized. In the high-rainfall zone the exponents added up exactly to 1.

The apparently high returns per unit of input in the low-rainfall zone reflect not only the necessary remuneration for the grazier's enterprise and management, and the generally increasing returns to scale, but also, Duloy points out, the great risks which the grazier has to bear, primarily risk of drought, but also uncertainties about his tenure and limitations on his stocking imposed by the State Government, of which most graziers in this zone are tenants.

TABLE 9

	Low-rainfall zone			High-rainfall zone		
	Inputs, £/year		Marginal product as multiple of present input	Inputs, £/year		Marginal product as multiple of present input
	Actual (geometric mean)	Optimal		Actual (geometric mean)	Optimal	
Land	1474	1200	2.07	1184	1177	1.41
Labour	1109	1082	2.48	941	713	1.08
Fencing[b]	514	813	4.01	270	444	2.34
Improved pasture[c]				215	277	1.83
Watering facilities[b]	351	642	4.64			
Machinery[d]	768	479	1.58			

[a] Six per cent on unimproved capital value.
[b] Depreciation, repairs and 6 per cent interest.
[c] Valued at 2 £/acre/year—in the opinion of some this should be £3.
[d] Repairs, fuel, and oil only. Depreciation and interest on machinery appeared to make no contribution to the production function, indicating previous over-investment.

Land use, however, is optimal in the high-rainfall zone, somewhat too high in the low-rainfall zone. Both zones should spend more on fencing; the high-rainfall zone should spend less on labour, the low-rainfall zone more on watering facilities and less on machinery.

A production function for very high-quality sheep land on the Canterbury Plains in New Zealand was prepared by Mason.[34]

TABLE 10

	Exponents	Marginal product at mean input, £	Price, £
Numerical factor in production function	3.08		
Land (acres)	0.42	3.1	25[a]
Superphosphate (cwt)	0.22	2.7	0.7
Lime (tons)	0.06	2.6	2.25
Labour,[b] £	0.15	0.85	1
Plant and machinery,[c] £	0.26[d]	0.7	1

[a] Purchase price—annual interest about £1.

[b] Including imputed labour of farmers, valued at current wage of 624£/year.

[c] Thirty per cent of capital value, representing 10 per cent depreciation, 5 per cent repairs, 5 per cent interest, 10 per cent other expenses.

[d] Capital value of plant and machinery used in production function.

The sum of the exponents, at 1.11, indicates increasing returns. The exponent for land is strikingly high. Farms should be larger, and use less labour and machinery and more superphosphate.

Another study[35] on New Zealand sheep, made earlier when wool prices were higher and costs lower, shows even greater marginal returns to land.

Acres were all converted to "adjusted acres" in light of the soil fertility. Returns were calculated after debiting imputed interest on working capital, and a managerial salary.

Though the figures (read from a diagram) are approximate, they bring out most strikingly the very high dependence of the marginal productivities of land and of labour on their relative abundance in comparison with each other—probably more than in other countries.

[34] *Australian Journal of Agricultural Economics*, Dec. 1960.
[35] Philpott, New Zealand Association of Economists, First Conference, 1958.

TABLE 11. *Marginal Productivities in New Zealand Sheep Farming, 1953–4*

Labour input, man-years/ 100 adjusted acres	0.2–0.4	0.4–0.6	0.6–0.8	0.8–1.0	1.0–1.2	1.2–1.6	1.6–2.0
Marginal productivity of labour, £/ man-year	1900	1200	1200	700	200	250	250
Marginal productivity of land, £/adjusted acre/year	5.5	5.5	7	7	12	12	16

This table shows that 125 acres/man of good land were required to provide the 1200£/man-year, which was about the effective wage (taking housing, etc., into account) prevailing in New Zealand in the 1950s. Even at the lowest labour inputs, however, the marginal productivity of the land remained unexpectedly high.

CHAPTER 4

LAND VALUES FROM PROGRAMMING

SO FAR we have dealt with mathematical production functions over a wide range of farm sizes. Marginal productivities from such functions plotted on diagrams inevitably give "smooth" relationships; and we are sometimes tempted to think that they may also be a little too smooth in the other sense of the word. Valuable though the Cobb–Douglas function has been, it is an important advance when we actually pro-gramme operations for various assumed areas of land, and other inputs. Marginal productivities estimated in this way are immediately dis-tinguishable on a diagram, because they always appear as a series of discrete changes at particular points, resembling an irregularly constructed staircase. The drawback to this method, however, is that programmes can only be prepared for a single farm, though we may make every effort to choose or design a fully representative farm; the Cobb–Douglas method enables us to collect more comprehensive but less accurate information about a variety of farm types.

One of the first and simplest examples of such a programme, which nevertheless should be of great interest in Asia, was a piece of common-sense designing, without mathematical programming, by Jewett for newly irrigated land in Iraq.[1]

If Jewett's estimates are correct, the possibility of marginal product rising with size of farm is interesting. There is evidence from other sources to show that farm family labour in Iraq is never fully occupied, even in the busy seasons of the year.

In African agriculture, Clayton[2] has pioneered programmes show-

[1] Government of Iraq Development Board, Diyala and Tigris Projects, Report No. 3, Hunting Technical Services, 1959.
[2] *Agrarian Development in Peasant Economies.*

TABLE 12. *Marginal Products in Iraq*

	Farm family with given labour force (conversion on 20 dinars/ton wheat)			
Size of holding, hectares	7.5	9.25	11.0	12.75
Size of arable	6.0	7.5	9.0	16.5
Product tons wheat equivalent gross	15.95	18.15	20.95	23.65
Product tons wheat equivalent net	10.05	11.75	14.0	16.45
Marginal product, tons/ hectare		0.97	1.28	1.40

ing how a farm of less than 3 hectares can keep one man and three women fully occupied throughout the year, earning a substantially larger income than the present average. But his studies only refer to a single farm and labour force, and cannot be used to estimate the marginal productivity of land.

Odero-Ogwell[3] has prepared a whole series of programmes for different zones in the Nyeri region of Kenya, taking sizes of farms and farm labour forces as he found them (Table 13). The marginal productivities of land have been expressed in terms of maize at the locally prevailing price of 15 £/ton. The large farm is over-supplied with land, while the labour inputs on the smaller farms are clearly over-optimal, as is shown by the exceptionally high marginal value of additional land to them. These facts alone do much to explain the recent history of Kenya.

We turn from "common-sense" budgets to those prepared by precise linear programming, requiring a large input of computer time, and still more of skill and thought, which are much scarcer commodities than computer time.

An excellent study has been undertaken in Cambridge.[4] "Common-sense" constraints were inserted to provide that not more than 20 per cent of the farm land should be under sugar beet and 10 per cent

---

[3] University of London Ph.D. Thesis, 1969.
[4] Cambridge Farm Economics Branch, Occasional Paper No. 9. The results have been read off the curves as published.

TABLE 13

Size of farm, hectares	0.5	1.2	1.6	2.2	3.0	3.7	4.8	6.3	32.5
Labour input adult male man-day equivalent/ha./year	1380	1138	850	620	455	563	565	427	83
Marginal productivity of land maize equivalent kg/ha./year	*	4465	4270	2296	1898	2296	2296	1898	0

* The produce of this farm was almost entirely for family consumption, and no marginal productivity was calculated.

under potatoes. The study referred to arable farms in East Anglia (Table 14). The marginal product of land is here measured net after meeting fixed costs, and in this respect is not comparable with the gross marginal products quoted in the previous chapter.

TABLE 14. *Net Marginal Product of Land, £/acre/year*

Acreage	30–50	50–70	70–100	100–120	120–150	150–200
Labour input in full-time man-equivalents 1	24.8	20.7	14.8	−6.1	−15.1	−25.1
1½	25.9	25.3	20.9	12.0	5.4	−9.9
2	25.9	25.3	27.9	15.0	12.9	6.5

The marginal product of land is highest on the smallest farms. The value of £25.9 corresponds to 2.4 tons wheat equivalent/hectare, much higher than on farms of corresponding size in Sweden. For a given size of farm, the marginal product of land rises when the labour supply is higher. The table also draws our attention to the important fact that the net marginal product can be negative. Some theoretical writers have failed to see this possibility. Farm land, after all, has substantial minimum maintenance requirements, and under certain circumstances, when the labour supply is over-strained, the marginal productivity of further land can thus be negative.

The very high figures in the first three columns indicate a situation in which the amount of land on the farm is insufficient to keep the given labour force occupied in the most economical manner. We should also remember that substantial inputs of equipment are concerned too and that the smaller farms must have certain "indivisibilities" in this respect, which means that their equipment cannot be fully occupied.

From the Cambridge figures it is also possible to calculate the marginal productivity of labour. On a 30-acre farm the marginal product of an additional man is literally zero. At the other end of the scale, if a 200-acre farm has a labour input of less than 2 man-years (a situation most unlikely to occur in fact) the marginal production of an additional man-year would be about £3000.

TABLE 15

Acreage		30	50	70	100	120	150	200
Marginal product of labour, £/man-year	Man-years							
	1–1½	0	42	226	600	1070	3022	3536
	1½–2	0	0	24	224	354	1214	2872

Much work in Cambridge has also been devoted to the precise planning[5] of a 200-acre farm in north-west Essex. The farm was fully provided with storage and drying facilities. Programmes were based on the assumption that use would be made wherever possible of the services of contractors, and also of casual labour. The latter point is of particular importance, because the farm was on such heavy soil

[5] F. G. Sturrock, private communications.

4*

that it was deemed necessary (at any rate, until recently) to employ hand labour for hoeing the sugar-beet. Common-sense constraints were applied to limit the amount of land under any one cereal to 100 acres, and under all cereals to 150 acres, and the number of sows to 15. At the time the programme was prepared the cost of one man-year was estimated at £550. On the available evidence, Sturrock thought

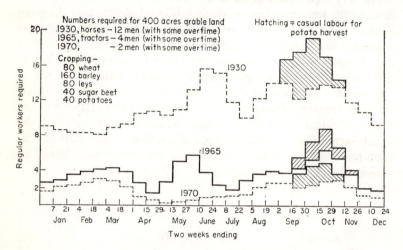

FIG. 7. Reduction in labour requirements due to mechanisation.

that the economic optimum for this type of farming might be a farm of 320 acres with three men and £9000 worth of equipment. He could envisage no further economies beyond this point.

There is a clear optimum labour input of 4 or 5 (40–50 acres/man). Below this, labour will not suffice to look after the farm properly, and marginal returns to labour will be very high.

However, Sturrock has plans,[6] in which labour will be still better equipped, and its optimal input lower. Subject to the availability of casual labour for potato harvesting, a farm of 400 acres which could be run by four men, with some overtime, in 1965 could, it was estimated, be run by two men, with some overtime (Fig. 7).

[6] *Advancement of Science* (British Association), Aug. 1966, p. 174.

TABLE 16. *Programme for 200-acre Farm in North-west Essex*

	Programmed						Present actual
Labour input, man-years	2	3	4	5	6	7	5
Gross product, £/acre	31	54	61	67	72	77	56
Net profit, £/acre	7.2	11.6	13.0	13.1	12.6	12.0	8.6
Marginal product of labour,* £/man-year	1430	830	570	450	430		

* Calculated from gross wage of £550 plus increment of net profits.

Maunder[7] devised an ingenious method of measuring the net marginal product of land under present-day conditions, by examining the accounts of farms which had recently increased or decreased their acreage, for periods before and after the change. In south-west England this was found to be 11 £/acre/year, in south-east England £16. In the south-east region "towards the top of the scale marginal product figures of £20–30 were not uncommon". This fairly clearly implies farms over-supplied with both labour and equipment in relation to their land resources. Such a situation was probably less frequent in the south-west. These marginal net products of land are, as we shall see shortly, far higher than the average product. If there is a gradual running down of the labour and equipment input on the smaller farms (by amalgamation or in other ways) the marginal product of land should fall to a figure nearer the average product. This, however, may take a long time

A copious and informative programming study for Sweden was undertaken by Hjelm.[8] Separate programmes were prepared for each region. Table 17 refers to the Southern Plains, the region with the best soil and climate, to a limited degree comparable with the eastern counties of England. In this case the marginal productivity of "working capital" (predominantly equipment, but including livestock and minor items) has been analysed out separately. The results are

[7] *Westminster Bank Review*, Nov. 1966.
[8] *Lantbrukhögskolans Annaler*, vol. 29, pp. 295–337, 1963.

expressed in money but converted to wheat equivalents on the 1961 average producer's price of 424 kronor/ton.

Some idea can be obtained of the marginal productivity of working capital, as defined, expressed as money net return/money input. Within the ranges of working capital utilisation programmed, marginal returns varied from 35 per cent to − 1 per cent with a median of 5.8 per cent as might perhaps have been expected.

To deduce marginal productivities of land, it is necessary to assume unchanged input of working capital, though in two cases alternative calculations are made on the assumption that working capital is increased or decreased at a 5.8 per cent marginal return (Table 17).

TABLE 17. *Marginal Product (wheat equivalent tons/hectare) on Farm Size Increments (hectares)*

Labour input, man-hours/yr.	Working capital input, tons wheat equiv.	15–30	30–60	60–90	90–120	120–150
3000	355	0.70[a]	1.29[a]			
5000	470		1.29[b]	0.63		1.04
7000	710				2.02	0.98

[a] The 30-hectare farm was over-supplied with working capital. Reducing this input to 235 tons wheat equivalent would have raised the 15–30-ha. figure to 1.03 and lowered the 30–60-ha. figure to 1.13.

[b] Reducing working capital on the 30-hectare farm as in note a would have reduced this figure to 0.82.

Some of these figures are unexpectedly high, with present labour and working capital inputs. Further increases in farm size would appear to be justified, if land could be obtained cheaply. The price of land reported by Sandqvist (see previous chapter) equivalent to 14 tons wheat equivalent/hectare for Southern Sweden appears to be about in equilibrium with the marginal product of land on the larger farms.

Raising labour input from 3000 to 5000 man-hours/farm/year shows marginal returns in the range 2½–7½ kg wheat equivalent/man-hour, according to size of farm and working capital input. But on the larger farms, other things being equal, raising labour input from 5000 to 7000 man-hours/farm/year shows a marginal return of 30–35 kg wheat equivalent/man-hour.

A thorough study was undertaken by Tyler[9] on the north-western slopes of New South Wales, in Australia, where land is abundant and labour scarce. In this region wheat-growing was clearly the most productive activity (wheat cultivation is now legally restricted throughout Australia) but sorghum, oats, lucerne, and the raising of beef cattle and of various strains of sheep were included as possible alternatives in the programmes.

Another programming study for wheat-growing in Australia was undertaken by Colin Hunt[10] in the low-rainfall Wimmera area of north-west Victoria. Production is now restricted by quotas; and some interesting programming studies were made about the exchange value of such quotas, and their effect on the price of land.

The programme is for a typical (though below optimum-sized) farm of 250 hectares. Dollar values are converted into wheat equivalents on the price at farm of 1.25 $A/bushel or 46 $A/ton (equivalent to 55 $U.S.). The programmes took account of fodder grains and livestock as alternatives. Alternatives are worked out for maximum and minimum expected prices for the principal fodder grain, barley—other fodder grains are expected to follow suit. Low fodder grain prices make it worth while to keep more sheep, thereby considerably increasing labour requirements. An allowance is made for additional managerial effort in operating a larger farm, but beyond that the farmer's labour is priced at the lowest unskilled rate ($1/hour). It will be observed that much of the farmer's potential labour time is unused (Table 18).

Tyler dealt mostly with much larger farms. On a one-man farm of 270 hectares his marginal product of 554 kg/hectare/year was much above Hunt's. The current price for land (carrying a reasonable quota) of 3250 kg/hectare (60 $A/acre) is compatible with the marginal returns on an average assumption of barley price.

---

[9] *Review of Marketing and Agricultural Economics*, Mar. 1964.
[10] Of South Australian Department of Agriculture. Australian Agricultural Economics Society 1972 Conference: and private communication.

TABLE 18. *Marginal Value of Land on a 250-hectare Wheat Farm (kg wheat equivalent/hectare)*

	Barley price 34.5$A/ton	Barley price 22.5$A/ton
Marginal return before debiting capital or labour but after debiting other inputs	629	376
Debits:		
Management (5 per cent of gross proceeds)	46	38
Cropping equipment (10 per cent depreciation plus 7 per cent interest)	38	32
Other equipment (do.)	22	22
Farmer's labour (1 $/hour)	88	121
Net marginal return	435	163
Farmers' labour unused hours/year	756	373

The main basis on which the programmes were framed were the alternatives of one, two, or three full-time labour units (very few farms in this region in fact had more than three). However, the programmes permitted the employment of a limited amount of casual labour for harvesting, and specialised contract labour for sheep shearing, crutching, etc. The farmers were assumed to be in good standing with their bankers and so able to obtain working capital without limit, for well-prepared programmes, at 6 per cent.

The permanent labour was regarded as being available for crop or livestock work only for 175 days per year, the rest of the available days being required for maintenance work, which could, however, be postponed during busy seasons. This labour limitation caused the marginal value of land for a one-man farm to drop suddenly to zero at 525 hectares. For two-man or three-man farms, on the other hand, marginal productivity showed a long gradual decline through very low values and not finally to disappear till 8100 hectares.

The determination of optimal sizes of farm depends on the rent payable for land. At the prices then prevailing of 40$A/acre and assuming 5 per cent return (in wheat terms[11] a price of only 1.91 tons/hectare

[11] Price received for wheat at farm 1.40 $A/bushel.

and a return of 95 kg/hectare/year) the optima were as shown in Table 19.

TABLE 19. *Optimal Area*
(hectares)

Men	Tractors 1	2	3
1	524		
2	548	585	
3		811	874

Marginal products (to land and labour) expressed in kg wheat were as in Table 20.

TABLE 20

One man, one tractor

Farm size, ha.	270	345	390	420	524						
MP, kg/ha.	554	391	344	210	206						
MP, kg/man-day		248	348	650	650						

Two men, two tractors

Farm size, ha.	490	573	579	585	2166	3490	8100				
MP, kg/ha.	563	391	344	210	38	29	7				
MP, kg/man-day		252	264	336	336	597	1110				

Three men, two tractors

Farm size, ha.	586	729	814	834	1067	1373	2165	2165	3260	7570	8100
MP, kg/ha.	563	362	162	91	71	54	49	38	37	29	7
MP, kg/man-day		248	248	255	349	340	340	365	596	1110	

The optima are marked by extremely sharp productivity falls, and would be much the same if the assumed rent for land were varied over quite a wide range. In contrast to some of the results shown by production functions, the marginal productivity of labour appears high. At that time casual harvest labour could be hired for the equivalent of 155 kg ($A8) per day. Separate calculations showed that on a three-man, two-tractor farm of optimum size of 811 hectares the marginal productivity of labour in the harvest season stood at the extraordinary figure of 2160 kg ($A112) per day.

For dairy farming we have a most interesting French study[12] by Hovelaque, which takes labour input in man-years/farm as given at a number of possible integers, and then programmes for each of these given labour inputs increasing land use, until the point of zero marginal return to land, i.e. optimal land use, is reached. He does not analyse inputs of equipment, but does distinguish the returns from three different types of farming, over a considerable range of farm sizes, with different assumed labour inputs. The study refers to north-western France, and the alternatives considered are dairying only (with incidental meat production), dairying with pigs, and finally a more labour-intensive dairying and pig production with zero grazing. Product is expressed in physical terms, of equivalents of milk as the principal product.

This is also a study of "programmed" productivities, with the labour supply in terms of integral numbers of men, i.e. not assuming the availability of part-time labour, as in the Cambridge study. Our first look at the diagram shows us with what remarkable uniformity this optimum works out at about 20 hectares/man (very similar to the English results) for labour inputs ranging between 2 and 6 man-years/ farm, and with three different types of farming. When approaching the optimum, land tends to have a slightly higher marginal value in the hands of the "dairying only" farmer than of the others. But on smaller farms this situation may be reversed.

Another point of interest is that we get one or two examples of the marginal productivity of land actually rising with increasing land input. Taking the wheat equivalent of a litre of milk at 1.3 kg, marginal returns to labour in this case did not rise above 67 tons wheat equivalent/man-year. They fell a quarter of this in the worst case, where four men were trying to make a living out of 25 hectares by dairying only.

[12] *Modèles des Structures d'Exploitations Agricoles*, École Agronomique, Rennes, pp. 90–94.

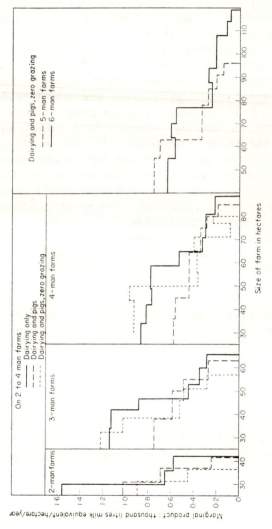

FIG. 8. Marginal productivities in France.

A study[13] for Bavaria is also available in which programmes have been prepared for a large number of farms, making careful distinctions between numerous soil and climatic zones. This leaves only a limited number (3–6) of farms in each zone; but in most cases an approximate linear fit can be made for the relationship income/labour unit to land/labour unit, indicating constant marginal returns to land. In other words, the range of densities covered, from 7 to 20 hectares/man, is still well below the optimum envisaged by the author for 1970–3, when all technical coefficients have continued to improve along their former trend lines, agricultural prices (in terms of other prices) are 10–15 per cent less than they were in 1960–3, but real wages have doubled in the course of the decade. "Income" is calculated after debiting all interest, maintenance and depreciation on equipment, and on new buildings, but not on old buildings or land. On the best arable land, in the neighbourhood of Würzburg and Regensburg, the marginal product of land was nearly 4 tons wheat equivalent/hectare/year, nearly as high as the average product. The median figure, however, was about 2.5 tons. A number of areas, including some hilly areas, showed no significant marginal product of land, and in the Allgäu, on the slopes of the mountains to the south, it was actually negative.

This review covers all the "programmed" marginal productivities which appear to be available, apart from a very thorough study by Kitsopanides of small farms in northern Greece. In this case also "short-term capital" was treated as an independent variable. On the larger farms the return on this was closely adjusted to the rate of interest, plus maintenance expenses, whereas on the smaller farms (which by definition could not afford all the capital they wanted) the marginal return might be ten times as high. In the case of land, the marginal returns on the larger farms (over 12 hectares) were surprisingly low, below 0.2 wheat equivalent tons/hectare/year, less than the prevailing rent of 0.33 unit. On the average-sized farm of 5 hectares, the marginal return was about 1 ton/hectare, rising to 5 on the smallest farms.

In the latter case almost the entire marginal product of the farm is attributable to the land. The marginal product of labour on the

[13] Plendl, *Stand und Entwicklungsmöglichkeiten Landwirtschaftlichen Betriebe in ausgewählten Naturainnen Bayerns*, Technischen Hochschule, München, 1967.

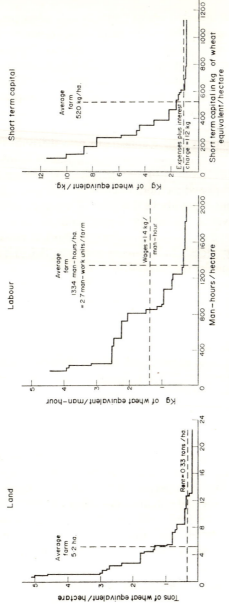

Fig. 9. Marginal productivities in northern Greece.[14]

[14] G. I. Kitsopanides, *Parametric Programming. An Application to Greek Farming*, University of Thessaloniki, Greece, 1965.

average farm, and indeed on most farms, is well below the wage actually paid (which implies that such a wage in fact is paid to only a limited number of men, employed in the busy season). The marginal product of labour on the farms where the man-hours/hectare are at their
highest (i.e. the smallest and most densely populated farms) falls to or
below the rate of 3 kg/day grain equivalent, which appears to be the
lowest rate of agricultural wages paid anywhere in the world, or at
any time in the historical past.

In this particular area there appears to be a sharp dividing line, at
about 800 man-hours/hectare, below which rate of labour input the
marginal return to labour is substantially higher. The optimum farm,
taking present crops and methods of cultivation as given, would therefore have a labour input above this level. If the average farm has $2\frac{1}{2}$
adult male equivalents of labour supply, and allowing for the fact
that there is a long cold winter in which no work is possible (one of
Professor Kitsopanides's programmes indicates that this would still
be the case even under the theoretical condition that capital was available in limitless quantities), we should regard the potential labour
supply as 5000 man-hours/year, calling for a farm of a little over 6
hectares. The average farm in fact is not far below this area, but the
distribution of farms by size is unequal.

Finally we may look at some land-use calculations for some of the
poorest, cheapest land in the world, namely the remote and arid cattle
properties in Queensland. The methods of calculation are crude, as
indeed are the pastoral methods of some of the cattle men in the re-
moter districts, which would not have looked unfamiliar to the patri-
archs of the Book of Genesis. The most important coefficients are
given by Beattie.[15] The information relates to cattle land, and predominantly wool-growing areas where the rainfall is a little higher, and
some improvements have been made on the properties, have been
omitted. Some simple algebraic transformations convert Beattie's
coefficients into expected annual beef output per unit of area.

The "value" of the land is ascertained in an unusual way. Almost
all of this land is state-owned, and is leased to graziers on long leases.
This arrangement suffers from the usual defect of state landowner-

[15] *Australian Commonwealth Scientific and Industrial Research Organisation.*
Bulletin No. 278.

ship, namely that it is too favourable to the tenants. However, we have some grounds for believing that the extent of these economic concessions is comparatively uniform between districts. The rents paid by the graziers are simply capitalised by the local authorities for the purpose of assessing rates (which in Queensland are assessed upon the assumed unimproved capital value of the land) (see Table 21).

TABLE 21

Information available	Proposed symbol
Total carrying capacity per sq. mile (mean of upper and lower figures)	p
Rearing percentage (mean of upper and lower figures) less 25 per cent because figures given refer to good seasons only	r
Average age of bullocks at slaughter (mean of upper and lower figures)	t
Average dressed weight of bullocks (mean of upper and lower figures)	w

Let $2x$ beasts (x of each sex) be reared per square mile per year.

Then by definition the number of breeding cows per square mile will be $2x/r$.

Assuming that they breed at the age of 2, the total cow and heifer population will then be $x(2+2/r)$. Male population will be xt. Then $p = x(t+2+2/r)$.

Total out-turn per year = total number bred

$$= 2x = p/[\tfrac{1}{2}t+(1+1/r)].$$

The above multiplied by w gives expected weight of beef produced per square mile per year.

Land with a capital value of 8 dollars per square mile is indeed about the cheapest land in the world. The value of the beef output, before paying transport costs, would be only 10–15 cents/lb at that time, dressed weight. Even after allowing for greater labour requirements in the more intensive producing areas, it appears that a surprisingly small proportion of the increased productivity was represented in higher land values.

TABLE 22

	p	r	$1+1/r$	t	Out-turn per sq. mile per year		Value of land ($A/sq. mile) (1959)
					Numbers,	lb beef	
Darling Downs and South-east (all Roma Division, Chincilla, Inglewood, Tara, Waggamba)	30	0.636	2.57	3.25	7.15	5190	842
Dawson Central Highlands and Beltando (Banana, Bauhinia, Belyando, Emerald, Jericho, Peak Downs, Tambo, Taroom)	1.75	0.525	2.90	3	3.98	2780	366
Isaacs and Mackenzie (Broadsound, Duaringa, Nebo)	20	0.506	2.97	3.75	4.12	3300	256
Boyne, Calliope and Fitz-roy (Fitzroy, Livingstone, Mt. Morgan)	28	0.580	2.72	3.75	6.09	4560	650
Upper Burdekin and Her-bert (Dalrymple, Mareeba)	17.5	0.337	3.96	5.5	2.60	1890	46
Lynd and Einasleigh (Etheridge)	13.5	0.337	3.96	5.5	2.01	1310	46
Lower Burdekin and Bowen (Ayr, Thuringowa, Wangaratta)	16	0.337	3.96	5.5	2.38	1750	138[c]
Mitchell Staaten and Gil-bert (Carpentaria, Croydon)	5.5	0.337	3.96	4[b]	0.92	370	20
Flinders, Leichhardt and Gregory (Barkly Tableland, Burke, Cloncurry)	11.5	0.337	3.96	5	1.78	1200	24
Cape York Peninsula (Cook)	3	0.25[a]	5.00	4[b]	0.43	170	8
S.W. Channel Country (Barcoo, Bulloo, Dia-mantina Quilpie)	4.5	0.25[a]	5.00	3.5	0.67	500	38

[a] Assumed—not given in Beattie's table.
[b] Assumed time required to breed store bullocks of 400 lb deadweight.
[c] Excluding Ayr Shire—predominantly agricultural.

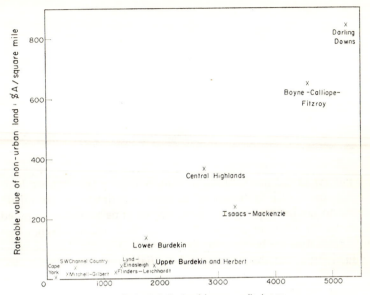

Estimated lb beef/square mile/year

FIG. 10. Productivity of Queensland.

The value of low-rainfall land (in this case in S.W. Queensland) for producing wool was shown by some experiments[16] at Gilruth Plains, Cunnamulla. At 1 hectare/sheep the yield was 4.08 kg of wool (greasy weight), raised to 4.51 kg/sheep at 2 hectares/sheep, indicating a marginal product of land of 0.43 kg greasy wool/hectare (a further increase of land input to 3 hectares/sheep raised the yield only to 4.56). This of course is minute compared with results for the marginal productivity of good-rainfall land.

For comparison with the previous table, the average product of 4.08 kg/hectare corresponds to 1180 lb/square mile of beef (dressed weight).

[16] Commonwealth Scientific and Industrial Research Organisation, 1941–2 Report.

CHAPTER 5

LAND RENTS AS A RESIDUAL

HAVING exhausted production functions and programming, we have another source of information in the form of estimates of "residual" income to land, i.e. gross product after debiting actual or imputed payments for all other inputs. The use of this method assumes that the other factors are being rewarded in accordance with their marginal productivities, that there are no substantial economies of scale, and that therefore the average returns to land should approximate to the marginal. We have already seen that in many circumstances this is not necessarily the case. However, it may be permissible to use this method in advanced economies, where a reasonable amount of factor mobility may be assumed.

To begin at low levels of productivity, we have estimates for a poor and very intensively farmed village in East Pakistan.[1] Here poverty

TABLE 23. *Residual Income in Sabilpur, 1961–2, rupees/hectare/year*

Gross product	968	Actual rent	526
Hired labour	142	Land taxes	21
Family labour[a]	363		
Other cash expenses	145		
Residual income	318		547

[a] Imputed at the rate of 1.2 R/day paid to hired labour.

[1] Habibullah, *Pattern of Agricultural Unemployment*, Dacca University.

and density of population appear to have led to a market very much out of equilibrium, with rent (including land taxes) taking two-thirds of factor income (gross product less cash expenses). This has the effects (i) of making net earnings per day of tenant farmers after rent less than half those of hired workers, (ii) of reducing the wages of hired workers themselves to below the marginal productivity of labour. In cases such as this it can genuinely be said that the owner of land is able to "exploit" the providers both of labour and of enterprise.

In India high residual values also prevail, particularly on small farms. C. P. Shastri found[2] in Western Uttar Pradesh (at prices of the early 1950s) results seen in Table 24.

TABLE 24

Farm size range in hectares	<1	1–2	2–3	3–4	4–6	6–8	>8
Relative frequency, %	3	14	15	14	22	15	17
Average area cultivated hectares	0.76	1.53	2.56	3.55	4.84	6.94	11.7
Gross income, R/hectare	2060	1499	1320	1330	1148	1200	1183
Residual[a] income, R/hectare	896	597	501	543	400	306	506

[a] Labour debited at 1.72 R/day.

The Institute of Agriculture at Anand[3] (Gujarat) conducted an interesting experiment in which four representative farmers were each given a tenancy of a 2-hectare block on condition that they purchased their inputs from the experimental farm, and kept full and accurate accounts (Table 25). In this case the rent is much *below* calculated residual income. The tenant farmer appears to be receiving a quasi-rent, probably at the expense of labour being paid well below its marginal productivity.

For the Nyeri District in Kenya Odero-Ogwel[4] gives a residual analysis as shown in Table 26.

[2] University of Agra Thesis, 1956.
[3] Bulletin 4, March 1958.
[4] University of London Ph.D. Thesis, 1969, pp. 247–9.

5*

TABLE 25. *Anand Farm Accounts, 1950–4 Average, rupees/hectare/year*

Gross sales	1821	Land tax	15
Bullock cultivation (hired)	147	Rent[c]	345
Hired labour	109		
Family labour[a]	231		
Water charges	214		
Seed	25		
Tractor cultivation	15		
Manure	119		
Plant protection	1		
Interest and depreciation[b]	54		
Residual income	906		360

[a] Imputed at hired labour rates of 1.56 R/day for men, 1.06 for women, 0.62 for boys.

[b] The farm's equipment cost 500 R, depreciated at 10 per cent per year, and other working capital 333 R. Interest was charged at 12 per cent per year on one-third of the equipment, i.e. it was assumed to be partly worn out, and on all the other working capital.

[c] Quoted as representative rent in this district by Desai, *Indian Journal of Agricultural Economics*, July 1961.

TABLE 26

	£/hectare/year
Product excluding crops grown for family subsistence and after meeting wage labour and other cash purchases of current farm inputs	64.7
Family labour (at opportunity cost)	34.7
Depreciation (at 5 per cent)	3.5
Interest (at 6 per cent)	2.7
Other overhead costs	2.1
Residual revenue to land	21.7
Tax	3.5
Residual revenue to landowner	18.2

We now turn to high-income farming, beginning with a comparative review[5] of different types of sheep farming in New Zealand.

The figures refer to predominantly sheep farms (Table 27). In comparing these with Philpott's figures given in the previous chapter it must be remembered they refer to a later period, of lower prices and higher costs.

Actual land prices (except in one region) were of the order of half the capitalised rents or less.

But when we consider[6] the country as a whole, including dairy farming, and for a more recent period, we find New Zealand farmers becoming more optimistic—strangely, in view of the gathering clouds over the world wool market, and the threat to New Zealand's dairy market implied by Britain's entry into the European Common Market.

Establishing wheat equivalents for land prices presents some difficulty, as wheat is not a significant part of New Zealand output. However, New Zealand prices are predominantly world market export prices, and are therefore linked to wheat on this basis. Over the period 1957–62 the f.o.b. export price of Australian wheat was equivalent to 41.2 $N.Z./ton. This was taken as a base, and the wheat equivalent of the New Zealand dollar for other years calculated from the New Zealand index of prices of farm output (Table 28).

Johnson (table 8) uses a residual income[7] after debiting maintenance and depreciation but before debiting rents, rates, land tax and interest, and after debiting an imputed 5 per cent on farmers' machinery and livestock. Johnson (table 13) debits for farmers' labour at the average rate of employees' remuneration. An additional debit has been made at the rate of 5 per cent of the gross product for managerial remuneration. Johnson points out that land prices change only one year after residual income changes with a marginal land price residual income ratio of 14.3.

[5] Keen and Gow, *New Zealand Meat and Wool Board Economic Service Bulletin*, No. 12, 1963.

[6] R. W. M. Johnson, Lincoln College, Agricultural Economics Research Unit, Technical Paper No. 4. Average prices have been weighted, to give representation to each farm size group.

[7] From official report *National Income and Expenditure* and Agricultural Economics Research Unit Research Report No. 59.

TABLE 27. Residual Income to Land in New Zealand Sheep Farming (mean of 3 years centred on 1959-60)

		S. Island highland	S. Island foothills	N. Island Hard Hill country	N. Island Hill country	S. Island fattening and breeding	S. Island intensive fat lamb	N. Island intensive fat lamb	S. Island mixed farming
Average farm size	Acres	29.983	2970	2762	1055	787	334	490	498
Structures (not dwelling)	£/acre	0.358	3.73	6.95	20.30	13.25	39.52	32.55	12.91
Equipment	£/acre	0.071	0.68	0.49	1.27	2.86	5.76	2.92	4.85
Livestock	£/acre	0.496	3.51	7.41	11.41	8.27	15.80	17.04	8.53
Labour	£/acre/year	0.087	0.333	1.05	1.26	1.06	1.88	2.31	1.85
Grazing and fodder expenses	£/acre/year	0.008	0.046	0.04	0.05	0.11	0.23	0.18	0.12
Other expenses including depreciation	£/acre/year	0.121	0.879	0.69	3.08	3.73	7.84	5.71	5.81
Net product[a]	£/acre/year	0.167	1.21	1.98	4.35	4.07	9.88	7.62	7.25
Economic rent[b]	£/acre/year	0.088	0.48	0.88	1.75	1.58	3.83	2.96	3.92
Do. receivable by owner[c]	£/acre/year	0.071	0.38	0.74	1.54	1.34	3.37	2.55	3.53
Do. capitalised[d]	£/acre	1.44	7.7	15.0	31.3	27.2	68.4	51.8	71.6
Land price	£/acre	0.53	5.56	4.28	8.85	15.3	23.4	27.0	34.4

[a] After expenses specified above.
[b] Net product after depreciation plus 5 per cent on capital items (shown above) and managerial effort imputed at 1000 £/farm/year.
[c] After deducting rates and land tax.
[d] At current bond yield of 4.92 per cent.

TABLE 28

	5 years ending 1957–8	3 years ending 1960–1	4 years ending 1964–5	4 years ending 1968–9
Land price, $N.Z./hectare	74.9	96.3	110.9	142.6
Do. wheat equivalent tons/hectare	1.66	2.34	2.59	3.19
Total residual income after managerial remuneration, $N.Z.m.	156.4	147.9	147.6	142.5
Land price/residual income	8.5	11.6	13.4	17.9

Good "residual" calculations can be made for Denmark.[8] The labour of the family is fully taken into account, including an allowance for managerial work "on the basis of salaries paid to hired managers" (which for these comparatively small farms come out at about 8 per cent of gross product, in contrast to the 5 per cent allowed below for the U.S.A. and the U.K.). Maintenance and depreciation of buildings, equipment and improvements are fully allowed for.

The tabulation in Table 29 differs from that published by the Danish Institute as follows: (i) taxation is excluded from costs,[9] (ii) an allowance is made for interest on all capital *not land and buildings* at a rate of $7\frac{1}{2}$ per cent. Residual income (less taxation) is capitalised on current yields on government bonds (Bank Rate before 1935).

Market value of land and buildings is shown (total market value of farm capital less book value of other assets). It always exceeds the capitalised value of residual income, but by very varying proportions.

The changing effect of expectations on land prices, in relation to capitalised rent, is seen more clearly from two series of data going back to 1910 for the United States.

The sudden rise of farm incomes during the First World War, farmers decided, was too good to last, so they bid down the price of

[8] Institute of Farm Management and Agricultural Economics (Landøkonomiske Driftsbureau). *Technical and Economic Changes in Danish Farming 1917–57*, pp. 72–74. *Thirty Years of Farm Accounts and Agricultural Economics in Denmark 1917–47*, p. 40. *Undersøgelser over landbrugets driftsforhold 1959–60*.

[9] In the years where only a combined figure for taxation and insurance is available taxation is assumed (on the basis of recent experience) to be three-quarters the sum.

TABLE 29. *Residual Incomes from Danish Agriculture*
kroner/hectare/year

	Five-year periods beginning:					Mean of 1958–9 and 1959–60
	1920–1	1925–6	1930–1	1935–6	1952–3	
Gross product	1053	811	563	643	2484	2880
Purchased feedingstuffs	229	204	121	115	430	823
Gross product net of feeding stuffs	824	607	442	428	2054	2057
Labour and management	347	279	216	248	910	971
Maintenance and depreciation of buildings and equipment	73	62	48	58	292	383
Interest at 7½ per cent on capital (not land or buildings)	57	48	36	35	103	168
Residual income (less other expenses)	172	66	30	83	232	146
Taxation (not deducted above)	36	33	24	30	94	109
Residual income as percentage of gross product net of purchased feeding stuffs	20.8	10.9	6.8	18.5	11.3	7.1
Value of land and buildings	2853	2617	2208	2310	3062	4246
Capitalised value of residual income less taxation	2165	614	154	1184	2565	1080

land in relation to current income from it. When prices suddenly fell in the early 1920s, on the other hand, they decided that this was too bad to last, and bid up the price of land in relation to its revenue. By the late 1920s this had given way to another mood of pessimism. Extremely low returns of the first half of the 1930s were regarded as transitory. But the later 1930s were again a pessimistic period. These alternating waves of optimism and pessimism can indeed be reduced

TABLE 30. *Residual Returns to Land, $ U.S. Billions*

	Gale Johnson[a]						United States Department of Agriculture[b]					
	1910–14	1915–19	1920–4	1925–9	1930–4	1935–9	1935–9	1945–9	1950–4	1955–9	1960–4	1965–6
Net income from agriculture[c]	5.20	8.98	7.28	8.32	4.57	6.28	6.36	18.61	18.29	16.05	17.54	21.20
Labour income[d]	3.27	4.38	5.12	4.90	3.12	3.93	4.24	11.68	11.23	10.07	9.26	9.28
Interest on working capital[e]	0.47	0.66	0.63	0.58	0.44	0.41	0.44	1.04	1.68	1.72	2.05	2.39
Allowance for managerial effort[f]	0.30	0.53	0.49	0.55	0.32	0.43	0.43	1.38	1.57	1.73	1.88	2.21
Residue (economic rent)	1.16	3.41	1.04	2.29	0.69	1.51	1.25	4.51	3.81	2.53	4.35	7.32
Bond yield, per cent[g]	4.29	4.93	5.46	4.69	4.53	3.26	3.26	2.65	1.41	1.97	2.93	3.43
Capitalised economic rent	27.0	69.3	19.0	48.9	15.3	46.3	38.3	170.0	264.4	128.4	148.7	213.4
Actual land price[h]	29.1	54.2	39.7	43.5	22.9	35.2	28.5	56.5	77.7	97.5	124.4	149.5
Land price/capitalised rent	1.08	0.78	2.09	0.89	1.50	0.76	0.75	0.33	0.29	0.26	0.84	0.70

[a] *Journal of Farm Economics*, 1948, p. 724.

[b] *Agricultural Finance Review*, June 1967.

[c] In the U.S.D.A. calculation exclusion is made of the imputed rental value of farmers' dwellings.

[d] Including imputed value of the labour of the farmer and his family.

[e] U.S.D.A.—at mean of bond and mortgage interest rates. Gale Johnson—at 6 per cent to 1934, 5½ per cent to 1937, subsequently 5 per cent.

[f] Taken at 5 per cent of *gross* product.

[g] AAA Bonds since 1918, previously railroad bonds. In the last three periods 1.73, 1.42 and 1.38 respectively were deducted to convert to "real" interest rate—see explanation in text. Data from *Review of Economics and Statistics*, Nov. 1970, p. 373. An arbitrary deduction of 1.5 was also made for 1950–4.

[h] Including farm buildings and fixed assets, except that U.S.D.A. excludes farm dwellings.

to an equation. G. T. Jones[10] has shown that current prices can be quite well explained as a product of the one-third power of current quinquennium's capitalised economic rents and of the two-thirds power of the prices of the previous quinquennium—and of actual current money prices at that; the land buyer appears to be extremely backward-looking, and does not even take into account general changes in the value of money.

In the later 1940s, surprising though it appears to us now, there was great pessimism about future prospects, and land prices were bid down to one-third of current earning capacity. In the late 1950s land prices rose rapidly. Whether buyers of land have been expecting further technical improvements, or further government support, we do not know. Land prices appear to have settled at about three-quarters of capitalised revenue. During the past decade net revenues have risen rapidly in spite of almost stationary prices for farm products through better technique and reduced costs.

Since 1965–6 there has been a remarkable rise in interest rates. These, however, should be considered as "money" rather than "real" interest rates—now that lenders have come to expect continual price rises, they expect thereby to lose part of their real assets by holding bonds—and bid up interest rates accordingly. The buyer of land, on the other hand, keeps his real asset intact, and therefore expects only a "real" interest rate.

Eckstein and Feldstein (*Review of Economics and Statistics*, November 1970) have constructed a system of distributed lags of past price changes (mean lag 8 quarters) and shown that this lagged weighted mean is almost fully reflected in current interest rates.

The Eckstein–Feldstein results were found to be applicable in the U.S.A. for the period since 1954. In the U.K., however, they do not appear to have applied before 1960. For earlier years no relationship between price movements and interest rates is discernible. Many marked rises and falls in price were experienced over this period; but for some reason investors did not expect them to continue.

In other countries also no adjustment has been made so far to

---

[10] Private communication. Actual equation (in logarithms) gives coefficients of 0.35 (0.07) and 0.71 (0.08) respectively. This equation, however, was computed on capitalisations which were based on money rather than on real rates of interest for the last four periods.

TABLE 31. *Economic Rents by Agricultural Regions, U.S.A.*

	Computed economic rent after charging imputed wages and mortgage interest rates on non-land assets, $/acre/year			% return on price of land		
	1937–41	1947–9	1957–61	1937–41	1947–9	1957–61
Cotton and other irrigated, California	—	54.9	55.8	—	16.3	7.3
Cotton (large scale), Mississippi Delta	—	13.7	17.8	—	16.3	11.0
Cotton, W. Texas	3.6	16.1	17.0	9.6	20.7	14.7
Tobacco (small scale), N. Carolina	—	15.6	15.1	—	11.6	7.0
Cash grains, Iowa	8.3	28.9	1.46	6.9	13.2	3.9
Wheat–pea, N.W.	5.3	20.0	13.9	8.0	11.4	5.2
Tobacco,cotton (large), N. Carolina	—	13.1	12.2	—	9.4	6.7
Hog–beef fattening, Corn Belt	7.4	34.5	11.3	9.3	22.6	4.5
Cotton (small scale), Mississippi Valley	—	18.7	9.3	—	23.3	5.7
Winter wheat(southern)	1.2	11.5	8.3	4.2	17.9	8.8
Wheat–fallow, N.W.	0.7	7.9	6.2	3.7	18.1	7.5
Cotton, E. Texas	3.8	10.3	5.3	7.4	12.0	3.6
Cotton, Piedmont	1.3	3.6	4.3	5.7	6.6	4.7
Dairying, N.E.	1.6	9.3	4.2	5.3	16.1	4.6
Mixed, N. Plains	0.3	5.9	0.8	1.5	21.5	1.7
Sheep, S.W.	0.28	0.29	0.7	4.9	2.9	3.8
Cattle, S.W.	0.12	0.39	0.38	2.9	4.6	3.6
Cattle, N.	0.04	0.74	0.21	1.1	8.4	1.9
Dairy–hog, S. Minnesota	0.4	3.8	− 0.4	0.7	3.9	− 0.9
Dairying, E. Wisconsin	− 2.2	− 5.2	− 10.4	− 2.7	− 4.4	− 6.0

"real" rates of interest, though it will probably be necessary from now onwards.

We should also observe, however, the remarkable regional differences which prevail in the U.S.A.[11] These figures do not contain an allowance for management, and should be reduced by about one-third if we want to compare them with the figures in Table 30. During the period 1957–61, to which it refers, the average money rate of interest was 3.85 per cent. But land prices were high in relation to capitalised revenue, and the average national yield on land was 2.7 per cent, not very different from the real rate of interest calculated above.

TABLE 32

Residual return	Zloty/ hectare/year
1926–9	81
1930–4	1
1934–7	18

Regionally, much the highest returns were on large-scale cotton growing, whether in the South or the West—does this mean that these growers are more than usually apprehensive about the future of their market? Tobacco showed a return of 7 per cent, and the principal Middle Western crops 4–7 per cent. Ranching showed much lower returns, perhaps indicating expectation of future appreciation of land values. Dairying (except in the North-east, where it was mainly for liquid milk sale) showed a negative return.

A "residual" calculation was attempted[12] for pre-Communist Poland debiting interest on all working capital and buildings provided by the tenant (no buildings provided by owners included in calculations). Considering that the purchasing power of the zloty was 2.2 kg

[11] Iden, *Agricultural Economic Research* (U.S. Department of Agriculture), April 1964.
[12] Deslarzes, *International Review of Agriculture*, July 1945, p. 111.

wheat in the first period, and about 4 kg in the 1930s, these are very
low returns.

For England and Wales, a long-period review of residual incomes is
possible. Residual income, or "economic rent", is defined from the
average gross produce of the land, less all material inputs, and less
all remuneration of factors of production other than the land and its
permanent structures. "Rent" in England and Wales is almost always
paid for land inclusive of these structures, and land prices are quoted
inclusive of them, though we must always bear in mind the fact they
cost money both to construct and to maintain. To estimate rent we
debit against gross product feeding stuffs, fertilisers, fuel, etc., purchased by the farmer, labour, an allowance for the farmer's efforts
both as manager and as manual worker, interest on the farmer's working capital, but not interest on any borrowings which the landowner
may have made, and depreciation and maintenance of the farm structures, but not interest on them.

Net factor income as now defined is comparable with the historic
estimates made by Bellerby,[13] subject to two differences of definition.
The first is a question of geographical coverage. The earlier data include the whole of Ireland, while post-1922 data refer to the present
boundaries of the United Kingdom. Ireland contributed approximately 25 per cent of the agricultural net output of the United Kingdom in both 1867 and 1908,[14] and it is assumed that the territory now
the Irish Republic contributed 20 per cent.

Bellerby made allowance for the valuation of self-supply by farm
families (relatively more important then than now). He did, however,
deduct rates levied by local authorities on farm land which, until
their final abolition in 1929, were payable by the tenant, not by the
landowner. On a precise definition, these rates must be included in
factor income, and they have been added back to Bellerby's figures.
Rates and similar charges must, however, be deducted from "economic rent receivable by the owner of the land", the variable which
we should compare with price of land.

Since 1945, most of this information on gross product and on inputs
can be obtained from the Ministry of Agriculture's *Annual Review*

---

[13] *Journal of Agricultural Economics*, Feb. 1953.
[14] See the present writer's *National Income and Outlay*, p. 246.

and Determination of Guarantees.[15] Adding back labour, rent and interest to "net income" there shown, we obtain a "gross factor income" comparable with Bellerby's.

The value of occupiers' capital is also given by Bellerby.[16] The Ministry of Agriculture's *Farm Management Survey* gives much lower figures, e.g. 27£/acre for 1954–5 as against Bellerby's 52£/acre for 1952–3. The principal difference is for crops and cultivations, which the Ministry value when they are at their lowest at the end of the year (5£/acre) and Bellerby at their highest (20£/acre). However, Bellerby appears to have also somewhat overvalued livestock and equipment. Up to 1952–3, therefore, Bellerby's figures are reduced by 40 per cent; subsequently the Ministry's figures are taken, with crop and cultivation values doubled. The rate of interest applied is the yield on government bonds plus 2½.

Regarding the allowance for management and farmer's labour, a figure of £5/acre (£2.6 for labour and £2.4 for management) was used by Wibberley when preparing his estimate of the true value of agricultural land at Lymm (Cheshire) in 1955–6, in connection with proposals for a new town on the site. The latter figure is very close to the 5 per cent of gross product taken as managerial allowance in the U.S.A. and this proportion of gross product is used throughout the table. The allowance for farmers' labour is extrapolated in proportion to agricultural wages.

The only available valuation of the existing stock of farm buildings and other fixed improvements in England is that made for two sample areas by Mr. H. J. Vaughan, a professional surveyor, for the Oxford

---

[15] In the *Annual Review 1964*, p. 28, retrospective figures of net income are given back to 1937–8, which are used in preference to those currently published. To these are added back currently published figures for rent and interest (this latter figure is assumed to have been rising £2 million annually from 1947–8 to 1951–2). If to the Ministry's net income estimates for 1937–8 we add the estimate by Raeburn (*Three Banks Review*, Dec. 1953) of £39 million gross rent, plus £5 million for interest, we obtain a factor income (excluding labour) for that year of £100 million, almost exactly the same as Raeburn's (Raeburn having, however, estimated labour income higher than Bellerby). For 1936–8 Bellerby estimates total factor income at £173 million, less labour at £115 million. Perhaps he did not allow sufficiently for rising industrial inputs. Adjustments of £10 million and £5 million respectively are made to his data for 1930–2 and 1923–9.

[16] *Journal of the Royal Statistical Society*, 1955, p. 340, for 1937–8 to 1952–3; for earlier years privately communicated.

Agricultural Economics Institute in 1960 (*Farm Economist*, 1962, No. 2, and Clark, "Capital requirements in agriculture", *Review of Income and Wealth*, Sept. 1967, p. 214). In the latter article, however, Thompson's valuation[17] of 1907 was erroneously quoted as *current* when in fact it represented *new* values; and therefore it should be approximately halved (on the assumption that structures were on the average at that time half-way through their working life). The depreciated net stock of farm structures (excluding farmers' and farm workers' dwellings) at 1960 prices was computed at £1122 million in 1907, £1248 million in 1948, £1240 million in 1960, and £1386 million at the end of 1965. (The agricultural rent restrictions in force in the 1950s discouraged owners from spending on their properties.)

Depreciation (excluding dwellings) in 1960 was calculated at £0.88/acre/year for one sample area and £1.04/acre/year for the other. Maintenance expenses[18] appear to have fallen in real terms (Table 33).

TABLE 33

	£/acre/year	
	Current	At 1960 prices
1938	0.45	1.56
1951	0.96	1.29
1957	0.92	0.95
1968	1.15	0.93

How far this fall has been due to tacit assumption by tenants of maintenance expenses previously borne by owners, and how far it represents genuine improvements (more economically designed buildings, ditching and post-hole machinery, etc.), it is difficult to say. Combined maintenance and depreciation on farm structures other than dwelling will be taken at £2/acre/year or £62 millions in 1960, and cal-

[17] *Journal of the Royal Statistical Society*, 1907.

[18] Dawe, *Country Landowners Association Proceedings*, 1953 (for 1938 and 1951); Bellerby, private communication (for 1957); Miles, *Journal of the Chartered Land Agents Society*, Jan. 1969 (for 1968). Bellerby specifically excluded maintenance of dwellings. It is not clear whether others have done so.

TABLE 34. *United Kingdom Agricultural Land—Residual Incomes*

		1867–73	1874–8	1879–83	1884–96	1897–1910	1911–14	1924–9	1930–2	1933–5	1936–8
Factor income[a] including Southern Ireland	£m/yr	155	150	122	115	116	137				
Factor income[b] present U.K. boundaries (Bellerby)	£m/yr	124	120	98	92	93	110	171	153	166	173
Labour costs[c]	£m/yr	45	46	41	38	37	40	65	60	57	58
Factor income less labour costs[c]	£m/yr	79	74	57	54	56	70	106	88	99	100
Interest on occupier's capital	£m/yr	16	16	14	12	12	14	23	18	14	15
Allowance for farmer's labour	£m/yr	7	8	8	8	8	9	17	16	16	17
Do. management	£m/yr	9	9	9	8	8	9	14	12	12	12
Maintenance and depreciation of structures	£m/yr	8	9	9	8	8	8	14	15	15	17
Residual income	£m/yr	39	32	17	18	20	30	38	27	43	39

	1947–8 to 1951–2	1952–3 to 1956–7	1957–8 to 1959–60	1960–1 to 1962–3	1963–4 to 1965–6
Factor income less labour costs[c] £m/year	353	417	449	536	588
Interest on occupier's capital £m/year	48	68	85	106	125
Allowance for farmer's labour £m/year	60	80	100	114	126
Do. management £m/year	48	69	79	87	96
Maintenance and depreciation of structures £m/year	45	58	62	66	77
Residual income £m/year	152	142	123	163	162

[a] Bellerby, *Journal of Agricultural Economics*, Feb. 1953. Includes changes in value of stocks and self-supply by farm families. Bellerby has debited rates, which are added back (data from Ojala, *Agriculture and Economic Progress*).

[b] Proportion of Southern Ireland to all U.K. found to be 24.3 per cent in both 1867 and 1908. Irish net output for 1867 and 1908 at 1913 prices (Clark, *National Income and Outlay*, p. 246), reduced by a further 5 per cent for other inputs, and 1867 data raised 5 per cent to bring from 1913 to current prices. Southern Ireland assumed to represent four-fifths of the whole.

[c] Since 1947–8 "Annual Review and Determination of Guarantees" with rent and interest added back to net income. Labour costs include national insurance, and estimate for unpaid labour of farm families, other than the farmer and his wife. Southern Irish wages before 1914 taken at 15 per cent of the whole. 1930–8, Bellerby adjusted (see footnote 18).

culated for other years on changes in net stock of structures shown above, and changes in non-residential building costs.[19]

For earlier periods, it is assumed that net stock was unchanged from 1880 to 1907, and rising 0.75 per cent per year before 1880.

Calculations of full economic rent before deductions for land tax, rates, and titles follow by subtraction.

The relation of this residual income to rents actually paid, and to land prices, will be considered in subsequent chapters. As these rents and land prices represent payments for use of farm buildings and improvements as well as land, including farmers' and farm workers' dwellings, their estimated net value after depreciation, at current prices, should be recorded. Improvements other than dwellings have

TABLE 35. *Net Value of Farm Improvements and Structures including Dwellings, £m*

	1960 prices	Current prices
1867–73	1720	270
1874–8	1760	294
1879–83	1800	288
1884–96	1800	258
1897–1910	1800	255
1911–14	1860	267
1924–9	1930	564
1930–2	1930	518
1933–5	1930	491
1936–8	1930	564
1947–8 to 1951–2	1930	1469
1952–3 to 1956–7	1930	1834
1957–8 to 1959–60	1930	1969
1960–1 to 1962–3	1980	2035
1963–4 to 1965–6	2060	2188

[19] *National Income and Expenditure*, and Redfern, *Journal of the Royal Statistical Society*, 1955, Part II, p. 170, for building costs.

been estimated above. Vaughan's results showed 21.9£/acre or £680 million in all for net value of dwellings, and this is assumed to have been unchanged throughout (in the nineteenth century "tied cottages" reckoned as part of the farm property were probably more numerous, but of lower quality).

CHAPTER 6

ACTUAL AGRICULTURAL RENTS

HAVING reviewed the information "explaining" (to an increasing degree of crudity) agricultural rents, it may now be of some value to assemble the available information, so far as is possible, including historical information, of rents actually paid in different countries and periods. Where they are expressed in money it is desirable to convert them, at current market price, to the unit of tons wheat equivalent/hectare/ year, which is independent of time and place.

Alternatively rent may be expressed as a proportion of the gross product, or of the factor income (in the simpler forms of agriculture these two differ little, if at all, but they differ greatly in advanced commercial agriculture with its large purchased inputs, e.g. by a ratio of over 2:1 in the U.S.A.). It is important also to consider taxes on land, about which information unfortunately is usually lacking. But it is probably safe to assume that most rents have been recorded gross of tax (i.e. that the tax has to be paid by the owner, not by the tenant).

It is also assumed that rents are gross of costs of maintenance, depreciation of permanent structures, ditches, etc., and of costs of collection and administration. In the simpler forms of agriculture these costs should not be very high, though for Egypt Bent Hansen[1] estimated them at 20 per cent of the gross rent. For the U.S.A. taxes and maintenance expenses were estimated[2] at one-third of gross rent.

---

[1] Private communication.
[2] *Farm Cost Situation*, May 1962, p. 20.

TABLE 36. *Rents (presumed gross of taxation, maintenance and admi-
nistrative expenses unless otherwise indicated)*

	Curren-cy unit	Rent/hectare/year		Rent as percent-age of	
		In money	In kg wheat equi-valent	Gross product	Factor income
AFRICA					
Egypt 1st century A.D.			638	34	
4th century A.D.			398	21	
1897	£E	10.4	1465	55	57
1923–7	£E	23.6	1875	52	55
1936–8	£E	15.0	1595	47	51
1947	£E	42.5	1985		
1950–1	£E	67.5	3150	44	48
1960	£E	48	2245		29
Ghana 1957				25–33	
Sudan 1952	£E	0.4		33	20
AMERICA					
Brazil 1944	$U.S.	5	200		
Chile 1960			600	44	
Surinam 1965–6			1400		
U.S.A.					
Iowa					
Indiana } 1962	$	47	630	20	42
Illinois					
Ohio					
Missouri					
Michigan } 1962	$	31	415	15.5	32
Minnesota					
Kansas					
Nebraska					
Dakotas } 1962	$	18.5	248	21.5	45
Wisconsin					
ASIA					
China average 1933				32	
Ting Hsien Irrigated 1928	$C	116	1025	35–40	
Not irrigated 1928	$C	65	575	35–40	
Yangtse Valley 1945				50	

TABLE 36 *(cont.)*

	Currency unit	Rent/hectare/year		Rent as percentage of	
		In money	In kg wheat equivalent	Gross product	Factor income
ASIA *(cont.)*					
India, Uttar Pradesh 1840	R	10.4	231		
1924	R	33.1	259		
1940–2	R	74.4	758		
1960	R	345	835		
Iraq North, unirrigated 1930–9					25
South, irrigated 1930–9					70–80
Japan 1838			765		
1878–87			1105	40	45
1888–97			1420	59	67
1898–1907			1468	68	78
1908–17			1530	63	75
1921			1445	58	72
1927			1580	51	67
1936	yen	311	1355	46	66
1938	yen	316	1580		
Malaysia 1948				53	
Krian 1928			328		
1954			606		
Kelantan 1955		230	673		
Kedah 1955			300		
Pakistan East 1961–2			625		
West 1960	R	547	821		
Philippines 1954–5	R	91	215	32	43
Syria irrigated 1930–9	pesos	163	570	38	41
not irrigated 1930–9	£S	325		60	
Syria 1965	£S	55–70		40–50	
Thailand			700	50	
Vietnam 1899			720	50	
1930			575	45	
1958			480	25	
EUROPE					
Belgium 1830					
1866	fr	57	160		
1963	fr	108	430		
France	fr	2700	565		
12th century				8	
Dauphiné Wheat 12th century				6–14	

TABLE 36 *(cont.)*

	Curren-cy unit	Rent/hectare/year		Rent as percent-age of	
		In money	In kg wheat equi-valent	Gross product	Factor income
EUROPE *(cont.)*					
France *(cont.)*					
Dauphiné Olive 12th century				9-25	
1300-50			94		
1350-1400	livre	8	69		
1450-1500			64		
1600-50	livre	17	65		
1650-1700			85		
1700-52		11	72		
1788	livre	26	170	35	40
1815	fr	35	150	22	
1845	fr	45	150	15	
1851	fr	50	240		
1879	fr	69	249	22	37
Best areas 1885	fr	132	450	15	
Worst areas 1885	fr	49	194	30	
1908	fr	57	245		
1938					22
1961			300		
Germany 1965	DM	380	900		
Greece 5th century B.C.			75		
Ireland 1884	£	1.36	180		19
N. Ireland 1925				14	
Italy 1st century A.D.	sesterces	235	730		
Italy North and Central					
8th-11th centuries A.D.				15	
12th century				40	
All Italy 1930-9				26	35
Average 1960	000 l.	22.5	320	14	19
Campania 1960	000 l.	50.5	720		
Basilicata 1960	000 l.	7.8	110		
Jugoslavia 1954	000 old	13.8	395		
(Voivodina)	dinars				
Netherlands 1923-9		7.8			30
1938		43		14	27
1940	guilder	74	615		
1957-62	guilder	1080	3625		
Romania 1886			260		
1905			300		

TABLE 36 *(cont.)*

	Curren-cy unit	Rent/hectare/year		Rent as percent-age of	
		In money	In kg wheat equi-valent	Gross product	Factor income
EUROPE *(cont.)*					
Spain 1903–12					40
Sweden 1954	Kr	120	280		
U.K. (England					
and Wales only 1376–1460	pence	20	150		
1688	pounds	1.05	124		37
1770	pounds	1.65	148		
1781–90	pounds	2.45	231		
1791–1800	pounds	2.62	178		
1801–10	pounds	2.72	142		
1811–20	pounds	2.84	141		
Arable land					
best	pounds	8.6	425	33	
good	pounds	4.1	203	25	71[a]
poor	pounds	2.1	104	20	
Grazing land					
best (Lincolnshire)	pounds	9.8	485	38	
normal				55	65[a]
1821–30	pounds	2.52	185		
1831–40	pounds	2.59	197		
1841–50	pounds	2.84	232		
1851–60	pounds	3.01	235		
1866–73	pounds	3.63	283		37
1874–8	pounds	3.85	336		40
1879–83	pounds	3.65	362		47
1884–96	pounds	3.21	380		44
1897–1910	pounds	2.84	413		38
1911–14	pounds	2.91	380		33
1923–9	pounds	3.88	358		29
1930–2	pounds	3.66	565		30
1933–5	pounds	3.31	645		25
1936–8	pounds	3.28	428		24
U.S.S.R.					
14th century				38	
16th century				33	
(Voronezh) 1903	rouble	15.3	220		

[a] Factor income defined *exclusive* of wages.

Sources and Notes to Table 36

Egypt   Early data from D'Avenel, *La Richesse Privée Depuis Sept Siècles*. Gross yield from Michell, *The Economics of Ancient Greece*. A. H. M. Jones in *The Later Roman Empire* stated that in Egypt rent (which was subject to tax) took half the crop: but on the Imperial estates in Africa in the second century it took one-third. From 1897 estimates by Bent Hansen (privately communicated) based on Willcocks (1897), Minost (1923–7), Anis (1936–8), Warriner, *Land Reform and Development in the Middle East*. Issawi, *Egypt at Mid Century*, and Ministry of Agriculture since 1947. 1960 rent from Eshag (private communication). Cultivated areas from El Kammash, *Farm Economist*, 1968. El Kammash quotes still higher rent 54.5£/hectare) for 1947. Gross product is measured net of agricultural inputs (seed and fodder). Taxation took 29 per cent of gross rent in 1897 (Issawi, *Egypt in Revolution*).

Ghana   Torto, Deputy Director of Agriculture, private communication. Maize growers on worn-out cocoa land, cropping twice a year.

Sudan   Versluys, *Economic Development and Cultural Change*, Oct. 1953. Refer to combined share of landowner and water-wheel owner, half being attributable to each, in the Gezira settlement. Ox power presumed provided by the tenant, otherwise charged at 20 per cent of gross product.

Brazil   (São Paulo) Foreign Agriculture (U.S.D.A.), June 1945.

Chile   Prof. T. Davis, private communication. Good irrigated land in Central Valley.

Surinam   Luning (see Chapter 3). This refers to the rents which farmers charge each other, not the low rent charged by the Government. Expressed in paddy, converted by a factor of 0.8. Marginal productivity in the same units 1925 kg.

United   *Farm Cost Situation*, May 1962, p. 21 (farms, not pasture land) and
States   *Agricultural Statistics of the United States*. Gross product per unit area computed on crop land (not under soil improvement crops or idle) plus half permanent pasture land.
Last column computed by applying national ratio of factor income to gross product.

China   Average from Shen-pao Nien-Chien, *Year Book*, Shanghai. Gamble, *Ting Hsien* (Institute of Pacific Relations), referring to a village near Peking. The rents represented 70–80 per cent of a single crop, but the land was mostly double cropped. For 1945, Hsiao-Jung Fei, *American Journal of Sociology*, July 1946.

India   Royal Commission on Agriculture 1926, Report on Agriculture in the United Provinces. Rents include Land Revenues of 4.0 R in 1840 and 6.4 R in 1924. Wheat prices from Brij Narain, *Indian Economic Life*, pp. 104, 108 (means of five years). Ghatge and Patel, *Indian Journal of Economics*, 1942–4, average of previous five years, for farms in Bombay Province, converted at prices of 1940–2. To these results have been added 5.9 R/hectare for land revenue (Gadgil, *Survey of Farm Business in Wai Taluka*, Gokhale Institute, 1940). Desai, *Indian Journal of Agricultural Economics*, July 1961, for land near Anand in Gujarat.

Iraq

Warriner, *Land and Poverty in the Middle East*. Landowners provide irrigation channels. Figures include respectively 5 and 10 for taxation.

Japan

1838: *International Institute of Agriculture Monthly Bulletin*, Oct. 1913. Refers to tax collected from peasants.
1878–1917: Tobata and Ohkawa (in Japanese) quoted Eicher and Witt, *Agriculture in Economic Development*, p. 54, last figure confirmed by *International Institute of Agriculture Monthly Bulletin*, July 1912, at 57 per cent for paddy land, 44 per cent for upland rice (latter not grown in substantial quantities).
Rice yields from Ogura and others, *Agricultural Development in Modern Japan*, Japan FAO Association, 1963, p. 682: yields stated as "husked rice" which is taken to represent 80 per cent extraction (Ishikawa, *Economic Development in Asian Perspective*), 1 kg of un-husked paddy expressed as wheat equivalent of 0.8 kg: rice yield in 1838 from Ando, National Research Institute of Agriculture, *Japanese History of Rice*, No. 3.
1921 and 1936 from Industrial Bank, quoted Dore, *Land Reform in Japan*, p. 43.
Tobata and Ohkawa describe their rents as levied on the single crop which he grows.
Dore, however, quoted 4 points higher (i.e. 55 and 50 per cent respectively) on double-cropped land.
Rents at a substantially higher proportion of gross product are quoted in *A Century of Technical Development in Japanese Agriculture* (1959), namely 62 per cent in 1921–7, 58 per cent in 1923–7, 55 per cent in 1928–32, 52 per cent in 1936. Ratio of factor income to gross product as given by Tsuru and Ohkawa, *International Association for Research in Income and Wealth*, vol. III, i.e. 0.88 for 1878 and earlier, 0.82 for 1906–13, 0.70 for 1933–7. 1927 rent from Yagi, *Kyoto University Economic Review*, 1930. 1938 rent from Singer, *Economic Record*, Dec. 1947. The original land tax of 1873 took 54.5 per cent of the gross product (15 per cent deducted for seed and fertiliser allowance valued at 16 years' purchase, then taxed at 3 per cent national plus 1 per cent local—see *I.I.A. Bulletin*, Oct. 1913), Rising prices had the effect of reducing tax, by 1912, to 16 per cent of gross product (*I.I.A. Bulletin*, July 1912). Rents quoted are inclusive of taxes.

Further alternative data (sources quoted in Chapter 7) are as follows:

	Rent per hectare paddy land Yen kg wheat equivalent	
1890	75	1110
1899	100	1150
1905–9	116	1015
1910–14	163	1425
1915–19	250	1170
1920–4	357	1255
1925–9	291	1155
1930–4	137	1110
1935–9	311	1150

Malaysia   For 1948, Institute for Medical Research, Kuala Lumpur, Report No. 13, 1950, p. 47. Second cropping not attempted because of livestock. For 1928, 1954 and 1955, Wilson, *Economics of Padi Production in N. Malaya*, Malayan Dept. of Agriculture, 1958. Kuchiba and Tsubouchi, *The Developing Economics* (Japan), Sept. 1967, quote a legal maximum rent of 870 kg/ha. paddy. This was sometimes but not often exceeded.

Pakistan   Habibullah, *Pattern of Agricultural Unemployment*, Dacca University.
East   Rent includes tax 21 R.
West   Khairpur Area, Naylor, *International Journal of Agrarian Affairs*, Oct. 1963, p. 11. Share rent including 27 R/ha. taxes.

Philippines *Central Experiment Station Bulletin* No. 1, 1957, pp. 106, 109. Share rent, after paying landowner's share of labour seed and other costs. Includes 9 pesos/ha. tax.

Syria   1930–9, Warriner, *Land and Poverty in the Middle East*, 1965. J. L. Simmons, D.Phil. Thesis, Bodleian Library, Oxford.

Thailand   Usher, private communication. Refers to rice land only—other land almost rent free. Land tax trivial.

Vietnam   R. L. Sansom, D.Phil. Thesis, Bodleian Library, Oxford. Hickey, *Village in Vietnam*, p. 45. Pre-1945 rents were chargeable on main crop only—but second cropping was then insignificant. 1958 rents refer to first and second crops combined, 1967 rents to first crop only.

Belgium   Leroy-Beaulieu, *La Répartition des Richesses* for 1830 and 1866. Bublot, *L'Exploitation Agricole*.

France   For 1885 Leroy-Beaulieu (see Belgium), for 1815 Chaptel (quoted Simiand, *Le Salaire*, vol. III, p. 95) and D'Avenel, *La Richesse Privée*

Depuis Sept Siècles for early data. For 1788, 1845 and 1885 values of production Mayer,*International Association for Research in Income and Wealth*, Series III, pp. 74, 82. Assumed area 40 million hectares cultivated. Livres converted to francs at par. Aggregate rental value of land including imputed rent of owner-occupied land as given by D'Avenel for 1788 and 1815 and Lavergne for 1845.

International Institute of Agriculture Bulletin of Agricultural Intelligence, May 1914, for rents in 1851, 1879 and 1908 (weighted average of arable, meadow and vineyard).

	francs/hectare		
	1851	1879	1908
Arable	42	57	54
Meadow	72	97	64
Vineyard	69	129	75

1938, Klatzmann, *Études et Conjoncture*, Dec. 1947.
1961, Chombart de Lauwe, private communication.
A remarkable series[a] for the Soissonais shows real rents almost constant since 1801, and a falling proportion of gross output.

	Rent francs/ hectare	Do. kg/wheat equivalent	Rent as per cent gross product
1801–20	63	252	31
1821–40	63	252	23
1841–60	69	249	18
1861–80	73	251	15
1881–90	58	252	13
1901–14	60	252	11
1937– 9	46[b]	193	8
1942–50	48[b]	229	8

Bonné[c] quotes the following figures of land rent as a percentage of agricultural income:

---

[a] Ferté, *Bulletin de la Société française d'Économie Rurale*, Apr. 1952.
[b] Gold francs.
[c] Quoted at International Symposium of Desert Research, Jerusalem 1952. Original source not disclosed.

1901	25.5
1916	24.0
1931	15.6
1947	9.0

Germany	Privately communicated. Refer to best available land.
Greece	Michell, *The Economics of Ancient Greece.*
Ireland	Giffen, *Economic Enquiries and Studies.* N. Ireland, Harkness, *Economic Journal*, 1929.
Italy	Herlihy, *Agricultural History*, April 1959; 1930–9, Gasparini, *Informazione* Svimez; 1960, *Annuario dell'Agricoltura Italiana.*
Jugoslavia	Private communication.
Netherlands	Van den Noort, Doctoral Thesis, Wageningen, 1965.
Romania	Mitrany, *The Land and the Peasant in Roumania.*
Spain	Vandellos, *Metron*, 1925. Factor income was estimated at 4.25 billion pesetas. Rent less taxes 37 per cent.
Sweden	*Jordbrukets Utredningsinstitutet Meddelande* 8 (1957).
U.K.	14th–15th century, Davenport, *The Economic Development of a Norfolk Manor.*
	1688, Gregory King, *Political Arithmetic.*
	1770, Caird, *English Agriculture in 1850–51.* Caird also estimated 3.31 for 1850 and 3.70 for 1878.
	1780–1820, Norton Trist and Gilbert, *Journal of the Royal Statistical Society*, 1891, p. 528 (originally published in *The Times*, 20 April 1889). Contemporary estimates by Board of Agriculture were higher, i.e. 2.17 for 1790, 2.99 for 1803 and 4.00 for 1813 (quoted Chambers and Mingay, *The Agricultural Revolution*).
	Data for 1817 from Sinclair, *The Code of Agriculture*, p. 45.
	1820–60, Thompson, *Journal of the Royal Statistical Society*, Dec. 1907. Data raised as Thompson suggests throughout. Thompson gives some pre-1816 data which were admittedly too low as Wales was over-represented. He quotes McCulloch's figures for this period, which also seem unduly low. Norton Trist and Gilbert give more fluctuating values as follows:

1821–30	1831–40	1841–50	1850–60	1861–70	1871–80
1.98	2.69	3.11	3.60	3.06	3.65

1866–1938, Bellerby, *Journal of Agricultural Economics*, Feb. 1953. Thompson's figures are almost identical.

Factor incomes from Bellerby, adjusted—see previous chapter.

U.S.S.R.	Chayanov, *The Theory of Peasant Economy* (translation published by American Economic Association, 1966) for 1903. Earlier data Blum, *Lord and Peasant in Russia*, pp. 102, 221.

From the ancient world we are fortunate in being able to obtain precise information from Egypt. The fall in rents between the first and fourth centuries A.D., whether measured in absolute amounts of grain, or as a percentage of gross product, is almost certain evidence of a declining population—for which we have other circumstantial evidence also, including a fall in the rents of houses.

In modern Egypt, population density is known to have been increasing rapidly. In the nineteenth century, when rural population density was much lower, there were indeed complaints of labour shortages. Nevertheless, the proportion of product taken in rent was falling gradually even before the land reforms of 1952, and heavily after that. This can only be explained by the increasing availability of industrial employment in Egypt.

It will be remembered that the period 1950–1 was one of exceptionally high prices for cotton.

For the United States information is available for some Middle Western states, which are combined into three groups, of declining soil and climatic qualifications. The national aggregate "residual income" of 3.35 billion average for 1960–4 computed in Table 36 must remain a national aggregate, and cannot reasonably be re-expressed per average hectare, because of these very wide differences in the quality of land.

There is a remarkable scarcity of data for India. One reason for this is that rents have been legally restricted by the State Governments, which acquired power to do this as long ago as the 1930s. The legal rents were always fixed well below the market equilibrium level. One consequence of this was that both land-owners and tenants became reluctant to give information about actual rents paid.

The figures for Iraq show clearly the effects of differing population density.

We have one figure for Tokugawa (pre-1868) in Japan, "whose economic measures displayed an amazing similarity to certain policies of eighteenth-century western enlightened despots"[3]—in spite of Japan's conscious effort to isolate itself completely from the rest of the world. In the legal sense of the word these were not rents. The Japanese peasant had become, for all practical purposes, the owner of the land which he cultivated, and his farming technique was reasonably advan-

[3] Kurt Singer, *Economic Record*, Dec. 1947.

ced. The payments by the peasants were legally described as taxation, and collected by the Samurai, or hereditary warrior class. This was the only form of taxation, apart from some sporadic taxation upon merchants. The tax was collected in kind.

After the restoration of imperial power by the Emperor Meiji—one of the most remarkable characters in world history—in 1868 began the thoroughgoing modernisation and Westernisation of almost every aspect of Japanese life. The privileges of the Samurai were abolished at one blow (though there were intermittent uprisings of deprived nobility and Samurai until the 1880s). In place of the revenues for- merly payable to them, the Emperor instituted a still larger tax payable to himself (and some to local authorities). This tax was to be payable in money, in the Western style. A highly rational valuation formula was enacted in 1873 (though accurate valuation was not in fact com- pleted until the late 1880s). For each plot of land the average rice crop was estimated, the price multiple applied, a deduction of 15 per cent allowed from the gross revenue to cover seed, fertiliser and other current expenses, the land then capitalised at 16 years' purchase of this net revenue, and a charge of 3 per cent per annum national taxa- tion plus 1 per cent per annum local made on this capitalised value.

The one thing which this intelligent and far-sighted emperor failed to foresee was the change in prices. World prices, in terms of gold cur- rencies, were due to fall for a quarter of a century after 1873. But Japan was on a silver currency, and the price of silver was falling heavily rela- tive to gold over this period. This had the effect of rapidly reducing the real burden of land taxation, in spite of temporary increases in tax rates at the time of the Sino–Japanese War of 1895 and the Russo–Japanese War of 1904–5. However, the very high rates of land taxation in the earlier period had done their work. They were a powerful force com- pelling Japanese farmers to increase productivity, and in also financ- ing education and industrial investment, as well as war prepara- tions.

The fall in the price of silver in terms of gold also had the effect of reducing the money labour cost (though not the real labour cost) of Japanese goods for export, and played an important part in getting Japanese industrial exports started as early as the 1890s. One may well speculate that, had the price of silver not fallen, Japan might have found it necessary to devalue the yen at that time.

These taxes were a declining element in gross rents, which were rising. With the virtual absence of industrial employment in the early decades, and the rapid growth of population in a country already densely inhabited, rents rose to an extraordinarily high proportion of total factor income, at its maximum in the 1890s.

It is true that in a densely populated community of independent peasants, as was nineteenth-century Japan, the renting of land does not play a large part. But the Emperor, pressed no doubt by Western economic advisers full of contemporary Ricardian theory, sought to augment the amount of tenancy. By an enactment permitting the investment of urban fortunes in the purchase of agricultural land, which had hitherto been prohibited, a class of large landowners in Japan was artificially created,[4] though at no time did they hold a large proportion of the whole agricultural land. These landowners remained until they were expropriated (with almost nominal compensation) by the American military administration after 1945.

It is surprising that rents should take such a high proportion of the gross product in Malaysia, a country with very low population density. The explanation which is given is that, although over 80 per cent of the area of the country is still uninhabited jungle, neither the Government, the landowners nor the customs of the villages permit this to be freely settled, and the inhabitants have to remain in the settled villages, paying high rents if they do not own land. In 1948, the year to which this information refers, "the Emergency" was in full progress, and many of the villages were deliberately fenced off from the jungle, in order to prevent Communist guerillas from obtaining supplies.

The rent in Thailand is a high proportion of the gross product but this is collected on rice only, other land being virtually rent free.

The figures for Vietnam are of unusual interest. Much of what is now densely populated land was still almost uninhabited as recently as the 1920s. However, mobility seems to have been fairly low, and rents in the established settlements had risen to as high as 50 per cent of the gross product in 1899.

Sansom's figures are qualified by Hickey,[5] who suggests a ratio of rent to gross product of only 40 per cent in the 1930s. Moreover, this

---

[4] Dore, *Land Reform in Japan*, Oxford University Press, 1959.
[5] *Village in Vietnam*, p. 45.

charge was only applicable to the main crop, though in fact second cropping was rare until the 1940s.

During this period of population growth and extensive settlement, average yields per hectare were falling slightly, as less favourable lands were brought under cultivation. The French administration had provided a good rail, road and water transport system by the 1930s, and a substantial part of the rice crop was exported, at the highly fluctuating prices prevailing in the world market. Nevertheless, there is evidence that during this period there was a substantial rise in the real wages of rural labourers, which were much higher than those in most of Asia.

The period of Japanese occupation, beginning in 1940, both reduced rents and encouraged second cropping (to provide vegetables for the Occupation Forces). Comparatively disturbed conditions (though not warfare on the scale of recent years) have prevailed over most of Vietnam ever since 1945. Many landowners did not attempt to collect their rents under these circumstances. However, economic as well as political factors seemed to have been at work in reducing rents. The figure of 25 per cent of gross product for 1958 refers to both first and second crops. By 1966/7 in the Mekong Delta, according to Sansom, rents were only 15 per cent of the first crop, and in the less militarily secure areas only 5–10 per cent. For one village, Sansom has provided a most interesting diagram, showing rents at an average of 20 per cent of the crop within 2 kilometres of the military guard post, but thereafter falling rapidly to zero at a distance of 5 kilometres.

In the Vietcong-controlled areas a carefully controlled system of taxation prevailed, taking between 20 per cent and 40 per cent of the produce of each family beyond a low per head subsistence allowance.

D'Avenel's studies on France are of exceptional interest. The low proportion of gross product taken by rent in the early Middle Ages shows that these were times of population sparsity, with landlords competing for tenants, whatever may have been the appearance to the contrary. The olive-growing country, however, appears to have been more densely settled.

However, the figures for rents in the first half of the fourteenth century, before the Great Plague, represent a higher proportion of gross production. We do not have a French estimate, but in England[6]

[6] M. K. Bennett, *Economic History*, vol. 3.

fourteenth-century yields were estimated at only 540 kg/hectare. The rent was thus 17½ per cent of the gross product, and a much higher proportion of the factor income, when we bear in mind that medieval methods of broadcasting seed (instead of our more economical practice of drilling) used as much as 170 kg/hectare. It is known that population density had greatly risen; and some estimates put the population of early-fourteenth-century France at as high as 30 million.

But the Great Plague, and the Hundred Years War which followed it, depopulated large areas of the country. After a respite, there was further devastation in the religious wars of the sixteenth century. The level of rents, which is good circumstantial evidence of population density, did not recover until the middle of the seventeenth century. This was followed by a further fall in the early eighteenth century, probably attributable to population losses in the prolonged wars fought by Louis XIV. In the later eighteenth century, before the Revolution, it seems clear that rents were rising rapidly, and played an important part in generating political upheaval.

In modern France there is little rented land, and information about rents is scanty. A special study of the Soissonais district,[7] a productive and technically advanced area, shows the real value of rents remaining quite unchanged at 250 kg/hectare of wheat equivalent from 1801 to 1914, although this represented 32 per cent of the gross product in the first two decades and only 11 per cent in the last. In 1942–50 rents were 229 kg/hectare or 8 per cent of the gross product.

For Italy also we have historical information going back to the early Middle Ages. Here the rise in population may have been no more rapid than in France; but the available land was probably used up more quickly, and by the twelfth century rents had risen to a high level. But we have abundant evidence, in the early Middle Ages, of landlords competing for tenants. A Church Council at Cremona in 1066 issued an edict defining fair commercial rents at that time—"let leases be given only for one-quarter of the grain and one-third of the wine". (This edict was necessary, not to protect tenants against landowners, but for the opposite purpose, namely to prevent the leasing of Church lands on beneficial tenancies to favoured occupiers.)

During the 1950s, with falling rural population density and in-

---

[7] Ferté, *Bulletin de la Société Française d'Economie Rurale*, Apr. 1952.

creasing industrial employment, there was a marked fall in real rents in Italy. Since 1960 this fall appears to have been checked.

In the Netherlands, strict government control on prices and rents of land has been in force ever since 1940. This, however, has provoked considerable evasion among an otherwise extremely law-abiding people. The estimates of rents actually paid during recent years—referring to clay land, which is above the national average, which is brought down by large areas of poor sandy land—nevertheless show an extreme rise, in real terms. Possibly the existence of the regulations causes the supply of land available for renting to be very limited, and the rents correspondingly high.

The figures for England and Wales are of great interest.

At first sight it appears that medieval rents were higher in real terms than those of the seventeenth century. However, the medieval information was for a manor in Norfolk, of productive arable land, and the average for the whole country, including grazing land, might have been considerably lower.

The rise in both money and real rents during the eighteenth century is apparent, accelerating after 1780 as a consequence of increasing rural population pressure. There appears, however, to have been a considerable immobility of rents, due to a high proportion of long leases, about that time, and the violent price increases after 1790 were not accompanied by corresponding rises in average money rents. In other words, much of the price increase accrued as a quasi-rent to the farmers—not to the wage workers, whose real wages promptly fell.

Before the railway age (see Chapter 2) the cities produced vast accumulations of horse and cow manure, which could only be carted for a minimum distance, thus creating a belt of exceptionally highly fertilised (and highly rented) market garden and dairying land in the immediate neighbourhood of the city.

Sinclair's rent figures for 1817 are quoted in Chapter 2. The gross rent of 20£/acre/year paid on 1400 acres of market-garden land near London was one-third of the gross output (a much higher proportion of factor income). On the 500 acres of such land near Edinburgh rent was 22–23 per cent gross product. Some vegetable land under glass frames had a product as high as 200£/acre/year. In addition, there were some plough-cultivated vegetable lands paying 3–5

7*

£/acre/year rent. Sinclair makes the interesting observation that "recently" these rents and land prices had fallen by 33–40 per cent — the consequences of improved road and canal transport? They were probably destined to fall further in the 1820s.

It is a well-known paradox of economic history that the Repeal of the Corn Laws in 1846, though bitterly contested at the time, was not in fact followed by a fall in real rents, or decline in the prosperity of landowners or farmers, the main reason being that world cereal prices were rising. Rent as a proportion of factor income in fact reached its maximum in 1879–83. Agricultural historians have succeeded in creating the impression of a heavy fall in real rents after 1879 (the very wet year which was taken as an historical turning point—many farmers were so disgusted by the weather that they voted Liberal for the first time, ensuring Gladstone's return in the election of 1880). In fact, as we see, the real value of rents continued to rise slightly though falling as a proportion of factor income. The trouble arose because landowners were trying to meet their own increasing standards of consumption out of an only slightly rising level of real rents.

From 1940 to 1958, and to a less extent since then, rents have been severely regulated by government action. An analysis of the figures would therefore be of little interest.

The information for Jugoslavia relates to the most prosperous agricultural region of the country. While the Communist Government permits owership and sale of agricultural land by working occupiers, renting of land is generally forbidden. An exception is made, however, in the case of co-operatives leasing land from small proprietors. In the case quoted, 6800 dinars/hectare, or nearly half the gross rent, represented taxation. The rate of tax was doubled shortly afterwards. The apparent downward trend in Russia is interesting—the 1903 figure represents only about 25 per cent of the gross product. The fourteenth-century figures refer only to the Novgorod region. Population density may have been falling as fresh lands were settled.

CHAPTER 7

ACTUAL LAND PRICES

To FACILITATE comparisons between different countries and different periods, land prices are expressed in units of tons wheat equivalent/ hectare. Regional data for a number of countries, showing the relationship between the price of land and rural population density (in the absence of alternative industrial employment), have been shown in Chapter 3.

Prices of land in relation to "residual incomes" capitalised at current rate of bond interest for national and regional data for New Zealand, Denmark and the U.S.A. are shown in Chapter 5. In the U.S.A., subject to considerable regional variations, price is found to be related to capitalised residual income, with long time lags. In Denmark, on the other hand, it was far higher. Rasmussen[1] examining data for a large number of individual farms over the period 1925–50 found

Price of farm end year/beginning year

$$= 0.923 + 2.89 \; \frac{\text{Net return}}{\text{Market value of farm}},$$

i.e. prices remained stable only when the net return was as low as 2.66 per cent. The average rate of return on commercial investment in Denmark throughout this period was far higher. Experience since 1950 seems to show a still greater discrepancy. We are compelled to conclude, therefore, that the difference between these two returns must represent the advantages of land owernship other than the direct monetary return from farming which prevailed in Denmark—the feeling of personal security, the sporting rights, the social position,

[1] Private communication.

possibly some taxation advantage. In a country like Britain, the social position and the sporting rights of landownership count for a good deal more, and in addition there are Death Duty concessions on agricultural estates; we should expect therefore the rate of return on agricultural land to be even lower than in Denmark.

But hardly anybody seems to understand the position, judging from the persistency of the complaints, from so many countries, that landowners are not getting a fair return on their investment. Switzerland is an outstanding example. The very high protected internal price for cereals has made the rents of land high and prices of land have been raised higher still because of the non-monetary advantages of landownership. For over fifty years now Swiss farmers have been complaining that they are not getting a fair return on their investment and demanding (and getting) still more protection.

On the other hand, we must not assume that these non-monetary advantages of landownership are constant at all times; nor that the ratio of the price of land to its rent is independent of commercial rates of interest. In some countries, the non-monetary advantages of holding land are small. At a given level of rent, moreover, and with the non-monetary advantages also given, the price of land will depend, not only upon changes in the rate of interest, but also upon the political outlook. Where landowners feel that their political future is quite secure, they may bid land up to very high prices.

The situation in England and Wales can be examined more closely, over a longer period. Satisfactory data on land prices and rents are not available for Scotland or Ireland.

The land prices, and also the actual rents quoted in Chapter 6, refer to England and Wales, whereas the "economic rents" are calculated for the United Kingdom. However, the relative contributions Northern Ireland and Scotland to factor income of United Kingdom agriculture appear to be much the same as their proportionate share of crop and grass land (excluding rough grazings); and therefore economic rent per acre of farm land in England and Wales should not be appreciably different from that calculated for the United Kingdom. It is true that both rents and land prices are much lower in Scotland than in England and Wales; but these differences appear to arise largely out of differences in Scots law and custom, including hereditary tenures.

TABLE 37. *Rents, Residual Incomes and Land Prices in England and Wales*

	1770	1781–90	1791–1800	1801–10	1811–20	1821–30	1831–40	1841–50	1851–60	1866–73	1874–8	1879–83	1884–96	1897–1910	1911–14	1923–9	1930–2	1933–5	1936–8	1948–9 to 1951–2	1952–3 to 1956–7	1957–8 to 1959–60	1960–1 to 1962–3	1963–4 to 1965–6
Actual gross rent[a] £/ha./year	1.65	2.54	2.62	2.72	2.84	2.52	2.59	2.54	3.01	3.63	3.85	3.65	3.21	2.84	2.91	3.88	3.66	3.31	3.28					
Deductions:																								
Maintenance and depreciation[b]	0.23	0.26	0.30	0.37	0.42	0.35	0.37	0.40	0.48	0.64	0.71	0.71	0.64	0.64	0.64	1.11	1.19	1.11	1.35					
Tithe and Land Tax[c]	0.50	0.50	0.62	0.87	0.89	0.59	0.57	0.55	0.55	0.52	0.54	0.52	0.42	0.35	0.37	0.40	0.37	0.37	0.32					
Net rent receivable by owner	0.92	1.78	1.70	1.48	1.53	1.58	1.65	1.59	1.98	2.47	2.60	2.42	2.15	1.75	1.90	2.37	2.10	1.83	1.61					
Residual income[a] £/ha./year										3.11	2.55	1.35	1.43	1.59	2.39	3.03	2.15	3.43	3.11	12.10	11.31	9.80	13.00	12.90
Do. receivable by owner[d]										2.01	1.52	0.38	0.60	0.94	1.75	2.39	1.83	3.11	2.95	12.10	11.31	9.80	13.00	12.90
Capitalised on Government Bond Yields:																								
Net rent receivable by owner £/ha.	25.9	39.3	36.6	30.2	33.8	43.6	48.7	57.0	62.8	77.2	82.2	81.0	76.2	67.0	57.9	58.0	58.9	58.9	50.9					
Residual income receivable by owner £/ha.										62.3	48.0	12.7	21.3	35.9	53.4	43.6	43.6	100.2	93.2	349	265	218	276[f]	292[f]
Actual land price £/ha.	52.8[e]	88.8	81.1	87.8	91.0	55.6	62.8	85.0	94.2	112.0	128.4	94.0	61.8	50.2	56.8	69.5	56.6	63.3	67.8	196	188	234	333	544
Of which farm improvements and dwellings[g] £/ha.										21.5	23.4	22.9	20.5	20.3	21.3	44.9	41.3	39.1	44.9	117	145	157	162	174
Percentage of land price[h] represented by "amenity and expectation" factors:																								
On basis of actual rents	51	56	55	66	63	22	22	33	33	31	36	14	−23	−33	−2	17	13	7	25					
On basis of residual income										44	63	86	65	28	6	23	23	−58	−38	−78	−41	7	17	47

[a] Source given in Chapter 6.

[b] From 1866 as given in Chapter 5. Extrapolated back in proportion to wages. Sinclair *(The Code of Agriculture)*, 1817, gives repair costs at that time as 10 per cent of gross rents. Allowing for depreciation as well, we obtain 15 per cent as shown here.

[c] Land Tax stood at 2.2 £/ha./year in the 1860s (Ojala, *Agriculture and Economic Progress*) after which it was gradually converted. Tithes were only payable in England and Wales, but we will use the per-acre figure for England and Wales, because our land price data refer to England and Wales. Tithes are extrapolated back on the price of wheat. This method receives some confirmation from the estimates of the Board of Agriculture (quoted Chambers and Mingay, *The Agricultural Revolution 1750–1880*, pp. 85, 118, showing 0.5 for 1790 and 0.97 for 1813). Local rates, which were substantial in the nineteenth century, were payable by the *occupier*.

[d] i.e. less rates, Land Tax and tithe. Even though rates were payable by the occupier, it is considered that they were fully "passed on" to the landowner, i.e. reduced the amount of rent which a farmer was willing and able to pay. Data from Bellerby and Ojala. (Maintenance has already been deducted.)

[e] Based on contemporary estimates of number of years' purchase of gross rent. Thompson (*Oxford Economic Papers*, Oct. 1957) quoted Arthur Young, *An Enquiry into the Progressive Value of Money in England* (1812), who gives estimates for 1768–73, 1778–89, 1792–9, and 1805–11 of 32, 23·25, 27 and 28 respectively. For the latter period Thompson substitutes his own figure of 35 for Young's 28, and gives further estimates for 1830–4 and 1865–74 of 27 and 37.5 Except for 1770, however, the Norton–Trist–Gilbert values are used till Ward's figures become available.

[f] For these last two periods a "real" rate of interest is obtained (see Chapter 5) by deducting 1.2 and 2.75 respectively.

[g] See Table 35.

[h] i.e. full price without deducting for dwellings and improvements. It must be borne in mind (see text) that this difference also contains a housing factor when we are comparing land price with capitalised residual income, though not when we are comparing it with capitalised rents.

Available information on average prices as shown in recorded sales of agricultural land in England and Wales has been regularly assembled by G. H. Peters and A. H. Maunder.[2] With the protected tenancies of the last thirty years, land with vacant possession sells at substantially higher prices than land without, and it is the former which is recorded. The Inland Revenue are now publishing[3] information on average prices of farms sold with vacant possession—how the Government resisted previous demands for the publication of such useful and harmless information for sixty years or more is one of the marvels of bureaucracy. Their figures show that partial results (based on recorded auction sales) previously published in *Estates Gazette* and *Farmer and Stockbreeder* were some 30 per cent too high. From 1968 to 1970 the Inland Revenue recorded prices were stable at 530£/ha (214£/acre).

Broadly comparable information for the period 1918–36 is given by Britton[4] (not, however, distinguishing vacant possession sales). For the decades prior to 1918 we have a remarkable annual series[5] on rents and prices of agricultural land for the whole period 1781 to 1880. The data were compiled from their own records by a long-established firm of land agents, Messrs. Norton, Trist and Gilbert. For some individual years in the earlier part of the period, the amount of land for which the averages were calculated was small, but for most years it appears to be adequate. Their figures apply only to agricultural land properly so called (with the buildings and other fixed structures of course included), excluding heath land and rough grazings, holdings below 30 acres, and building land. Norton, Trist and Gilbert also pointed out that, in the period which they are reviewing, most estates over 6000 acres would have been entailed, and would never have appeared on the market. In fact, the breaking of a number of ancient entails was made possible by the Settlements of Lands Act, 1882. This may have caused substantial additional quantities of land

---

[2] Published at intervals in *Farm Economist* and *Estates Gazette* (10 Apr. 1971, and corresponding earlier issues).

[3] Ministry of Agriculture, *Agricultural Land Service Technical Report*, 20 Mar. 1970.

[4] *Farm Economist*, 1949.

[5] *Journal of the Royal Statistical Society*, 1891, p. 528—originally published in *The Times*, 20 Apr. 1889.

to be put on the market, and contributed to the fall[6] in land prices which was taking place in the 1880s, though it was probably not a major factor.

J. T. Ward[7] gives a series in triennial groups, beginning in 1857, of farm prices recorded in the *Estates Gazette*, based on an average of some 250 sales per year. For years since 1918 his results are coordinated with those of Britton and Peters. Where Ward's data overlap with those of Norton, Trist and Gilbert the agreement is close. "Residual income receivable by owner" is based on the calculations in Chapter 6. It represents, in effect, the expected residual income to the owner if he farmed the land himself, at the standards of efficiency of the average farmer, paying a manager, and after meeting all rates, land taxes, tithes, maintenance and depreciation.

"The rent is not the full measure of the benefits of land holding", wrote Norton, Trist and Gilbert in their stately Victorian prose, "and that they have fetched these large number of years' purchase in open competition in the public market shows that the amenities attached to land were much appreciated." This "difference" represented a valuation of the political power, social prestige, sporting rights, etc., conveyed by landownership, and perhaps some element also of expectation of further increases. For the earlier decades it stood at well over half the price of the land. In the Napoleonic War period of high commodity prices the capitalised value of rents actually fell (interest rates were rising, and many rents were fixed on long leases). An "expectation" element appeared in land prices, though they did not rise significantly above the level of the 1780s. Land values came down to earth with a bump in the financial crisis of 1825. But then they recovered, in spite of the Reform Bill of 1832, which took away much of the political power of the landowners, and in spite of the chill winds of approaching Free Trade and Chartism. The "expectation" element in land prices increased. The threat of further drastic social changes had passed and landowners and potential landowners seeking a dignified investment for large fortunes made in business concluded that things were not so

---

[6] In an earlier and more optimistic period, on the other hand, it was thought that such legislation would make land more attractive to buyers, "We see no reason to doubt that if the transfer of land were simplified, its value might be increased by five years' purchase (of rents)" (Caird, *English Agriculture in 1850–51*, p. 496).

[7] *Estates Gazette Centenary Supplement*, 3 May 1958.

bad after all. Rents had risen, in spite of Free Trade, while the social amenities and political influence of landownership were still considerable.

When we compare the price of land with estimates of capitalised residual income—though not when we compare it with capitalised rents—we must bear in mind that the land includes housing for the farmer and for some farm workers, which has not played a direct part in producing residual income. In calculating residual income, the remuneration allowed for farmers and farm workers is *gross* of their housing requirements.

In the period of low prices between 1880 and 1910, the price of land fell below the capitalised value of actual rents, though it remained very high in relation to true economic rents—in other words, the landowners were grossly overcharging the farmers. From 1910 to 1932 land prices continued to contain an "amenity housing and expectation" element, though less than previously. After 1933, economic rents rose rapidly, while actual rents continued to fall. Land prices, however, did not rise in sympathy with the rise in economic rents—in other words, landowners, for some reason, did not expect the improvement to last.

Likewise the high economic rents of the period 1947–57 were far from being reflected in land values—investors probably regarded both high agricultural prices and low interest rates as impermanent. In any case, the landowner now had to farm the land himself if he wanted to get the economic rent from it, as rents paid by tenants were fixed at a very low level by government action, and in addition tenants were made virtually irremovable.

Since 1957 these legislative restrictions have been modified, though not removed, and economic rents have risen. But this rise has been to some extent offset by the rise in the rate of interest, even the "real" rate of interest. The price of land, however, has risen to a record height.

One element contributing to the difference between actual land prices and capitalised rents is said to be the mitigation of death duties on agricultural land (introduced in 1925). I am much indebted to Prof. G. H. Peters, who has prepared calculations of the extent of this effect on two assumptions:

A. Purchasers of land are assumed to be aged 50 with an expectation of life of 25 years who anticipate prices of land and interest rates at time of death to be equal to those at time of purchase. Purchasers are considered to be buying 150 acres and to have a total fortune subject to death duties of £50,000 in 1963/4–1965/6 extrapolated back for earlier periods in proportion to money GNP per head of population.

B.   Under this assumption the fortune in 1963/4–1965/6 was placed at £250,000 with 400 acres being purchased.

Given these assumptions, and knowing at each date the basic rates of death duty (from Reports of the Commissioners of H.M. Inland Revenue) and the abatement rate, plus the rate of interest on Consols, it is possible to calculate the present discounted value of the ultimate death duty concessions as shown in Table 38.

TABLE 38. *Capitalised Value of Death Duty Concessions, £/acre*

	1936–8 to 1946–7	1947–8 to 1951–2	1952–3 to 1956–7	1957–8 to 1959–60	1960–1 to 1962–3	1963–4 to 1965–6
Assumption A	–	3	3	4	5	9
Assumption B	2	9	7	9	10	18

For the earlier periods the value was inappreciable. It is clear also that the value of this concession is unlikely to contribute any substantial element to the present price of land.

Next we examine all other countries where the price of land can be compared with actual (or economic) rents (Table 39).

We may express price as a multiple of rent, or rent as a supposed rate of return on price. The former procedure has been chosen deliberately, in order to remind the reader once again of the important principle that the price of land is a *consequence* of its rent, and that to expect a rent as a consequence of applying a supposed rate of interest to the price of land is a fallacious form of reasoning.

It must be remembered in the first instance that all these rents, including the few estimates of economic rent, or of marginal product of

TABLE 39

		Price of hectare of land		Price as multiple of	
	Currency unit	In money	In tons wheat equivalent	Actual rent	Economic rent (residual income)
Egypt 1947–8	£E	1062	47.2	24.9	
Kenya (Nyeri dist.) 1967	£E	210	10.5		11.5
Sudan 1947	£E	34		8.0	
Chile 1960–1			10.0	16.7	
U.S.A. Iowa, Indiana, Illinois 1962	$	740	9.9	15.8	
Ohio, Missouri, Michigan, Minnesota, Wisconsin 1962	$	443	5.9	14.3	
Kansas, Nebraska, Dakotas 1962	$	253	3.4	13.7	
China Ting Hsien irrigated 1928	$C	1160	10.2	10.0	
Ting Hsien not irrigated 1928	$C	645	5.7	10.0	
India 1940–2	R	1090	11.1	15.8	
1950–4	R	3700	10.6	10.3	4.1
1954	R	5840	17.5		32.5
1960	R	2960	7.1	8.6	
Japan 1890	yen	605	8.9	8.0	
1899	yen	1270	14.6	12.7	
1905–9	yen	1502	13.1	12.9	
1910–14	yen	2520	22.0	15.4	
1915–19	yen	3652	17.1	14.6	
1920–4	yen	5470	19.2	15.3	
1925–9	yen	5050	20.0	17.3	
1930–3	yen	3790	22.6	20.4	
1935–9	yen	4480	16.6	14.4	
1950	000 yen	198	4.1		
1955	000 yen	1120	12.8		
1960	000 yen	1810	20.6		
1961–3	000 yen	1890	18.8		
Malaysia 1948	$M	466	2.0	6.1	
Krian Province 1928	$M		6.7	10.0	
Krian Province 1954	$M	2980	8.7	12.9	

TABLE 39 *(cont.)*

		Currency unit	Price of hectare of land		Price as multiple of	
			In money	In tons wheat equivalent	Actual rent	Economic rent (residual income)
Pakistan East	1961–2	R	16,100	24.1	29.5	50.5
West	1960	R	3030	7.2	33.3	
Philippines	1954–5	peso	2934	10.3	18.0	
Syria	1965	£S	1250	5.6	8.0	
Thailand		baht			12.5	
Vietnam	1899	piaster		4.4	4.9	
	1930			8.5	10.6	
	1958			8.0	13.3	
	1967			4.0	8.0	
Belgium	1963	000 fr	221	47.0		13.0
France 14th century		livre	130	1.06	13.2	
	1575	livre	279	1.16	17.9	
	1600–50	livre	292	1.35	20.7	
	1650–75	livre	481	2.26	26.5	
	1790	franc	764	4.36	25.8	
Arable	1851	franc	1470	6.6	35.0	
Pasture	1851	franc	2250	10.8	31.2	
Vineyard	1851	franc	2050	9.8	29.6	
Average	1851	franc	1665	8.0	33.3	
Arable	1879	franc	2060	7.5	36.2	
Pasture	1879	franc	2950	10.7	30.5	
Vineyard	1879	franc	2950	10.7	22.8	
Average	1879	franc	2290	8.3	33.2	
Arable	1908	franc	1485	6.8	27.5	
Pasture	1908	franc	1865	8.6	29.1	
Vineyard	1908	franc	2030	9.3	27.0	
Average	1908	franc	1591	7.3	28.0	
	1961	N.F.	2500	6.7	22.2	
Denmark	1917–21	kr	1253	2.8		12.0
	1922–6	kr	1138	4.1		20.3
	1927–30	kr	1023	5.6		35.3
	1931–4	kr	908	8.7		113.0
	1935–9	kr	915	5.2		28.5
Italy	200 B.C.	sesterces	400	1.2		
	150 B.C.	sesterces	4000	12.2		
1st century B.C.		sesterces	4750	14.7		

TABLE 39 *(cont.)*

		Price of hectare of land		Price as multiple of	
	Currency unit	In money	In tons wheat equivalent	Actual rent	Economic rent (residual income)
Italy 1st century A.D.	sesterces	4000	12.2	16.7	
Later Empire					
(post-tax rent)					25.0
Average 1960	000 l.	475	7.0	21.1	
Campania 1960	000 l.	916	12.0	16.2	
Basilicata 1960	000 l.	164	2.4	21.0	
Germany 1965	DM	18,000	42.5	47.2	
Jugoslavia (Voivodina)					
1960	000 old dinar	150	4.3	10.9	
Netherlands 1940	g	2550	21.2	34.5	
Poland 1926–9	zl	2060	8.3	51.7	
Switzerland 1901–5	franc	2489			37.1
1906–13	franc	2391			26.5
1923–7	franc	2848			95.0
1928–30	franc	2892			50.0
1935–9	franc	2997		8.5	42.9

Sources (where not previously specified in Table 36)

Egypt   Warriner, *Land Reform and Economic Development in the Middle East.*

Kenya   Odero-Ogwel, University of London Thesis, 1969, pp. 247–9. Valuation is in terms of maize, which is valued at 0.75 wheat unit. Residual income is net of tax. See Chapter 5.

Sudan   Versluys, *Economic Development and Cultural Change*, Oct. 1953.

India   1950–4, Institute of Agriculture, Anand, Bulletin 4, Mar. 1958. Rent includes land tax.
1954, Hopper, *Journal of Farm Economics*, Aug. 1965. Ganges Valley. Marginal product of land calculated from production function, not residual income.
1960, Desai, *Indian Journal of Agricultural Economics*, July 1961. Gujarat (Anand).

Japan   *Long-term Economic Statistics of Japan since 1868*, Vol. 9, table 34, refer to paddy land, and Vol. 8, table 12. Koku of paddy equated to 129 kg wheat equivalent.

Malaysia	Refers to rice land.
Belgium	Bublot, *L'Exploitation Agricole*. Economic rent is marginal product from production function—price/rent ratio approximately similar for all regions. Labeau, *Cahiers Economiques de Bruxelles*, 1965, for price (1962).
Denmark	Deslarzes, *International Review of Agricultural Economics*, July 1945, p. 111. Land price excludes buildings and structures, and residual income excludes interest on tenants' capital and buildings.
Italy	Earliest centuries, Tenney Frank, *Economic Survey of Ancient Rome*, pp. 125–365. A. H. M. Jones, *The Later Roman Empire*, p. 767; Duncan Jones, *Papers of the British School at Rome*, 1965, pp. 224, 279. Duncan Jones states normal wheat price at 2–2½ sesterces/modius, i.e. 323 sesterces/ton. The jugerum is taken at 0.252 hectare.
Jugoslavia	Mihailovicz, World Population Conference 1954.
Poland	Pytkowski, *Warsaw Biometric and Statistical Laboratory*, vol. II, 1937. Economic rent is marginal return computed from production function.
Switzerland	Deslarzes, *International Review of Agricultural Economics*, July 1945.

land, are *gross* of both taxation and maintenance. The price of land, however, may be expected to depend on rents *net* of these charges. To some extent therefore all the multiples shown should be raised somewhat. In the simpler forms of agriculture, the maintenance charges for land and permanent structures will be low. The taxation falling on land varies very greatly between countries, and has only been recorded here in a few cases.

In the ancient world, prices and rents in Italy were far higher than in Egypt. Duncan Jones comments on this, and points out the high interest rates and low liquidity of the capital market in Egypt. But another factor may have been that Italian land was still untaxed in the first century (A. H. M. Jones, p. 822).

The extremely low price in 200 B.C. reflects the utter devastation and depopulation of rural Italy as a result of Hannibal's invasion. The *latifundia* established in this era were cheaply bought.

A long series is available for France. The rise in the multiplier may be due in part to the fall in the rate of return on alternative investments, but alternatively may be due to a rise in the valuation of the amenity, social, sporting, etc., value of land. D'Avenel illustrates this by the remarkable fact that "le privilège de la chasse" only arose in the sixteenth century. Prior to that date hunting had been open to all. Indeed, in some districts there was a legal obligation upon landowners to hunt wild animals, whether they wished to or not.

But the other factor was at work as well. The rate of interest paid on loans secured on land fell from 10 per cent in the late fifteenth century to 6–7 per cent in the early seventeenth century. By 1789 the best secured loans, in some cases, were paying on $3\frac{1}{2}$ per cent interest. However, this does not explain the very high multiples of the mid-nineteenth century. At that time, rates of 4–5 per cent could be obtained on fully secured Government Bonds; and there was no expectation of a general inflation. The subsequent fall in the multiple can probably be explained by rising rates of interest. However, in 1961 French Government Bonds gave an average yield of 5.07 per cent. It is clear that land prices in France still include a large "amenity and expectation" element. The multiples for Switzerland are altogether exceptional. There appear to be very strong non-monetary reasons for holding land in that country.

Postan[8] records the surprising fact that in thirteenth-century England some land sales were recorded at forty years' purchase of the revenue. Possibly the ownership of these lands carried unusual social or political advantages. The contemporary rate of interest on loans secured on land might be anything between 5 per cent and 20 per cent.

The rise in the multiple in Japan shows (except for 1935–9) the fall in the real rate of interest and a better organised capital market, but also an increasing sense of political security. The Bank discount rate (rates on private loans were probably considerably higher) was 6.6 per cent in 1913, 8 per cent in 1921, and 5.9 per cent in 1927. It appears that there were few amenity or social elements in the price of land in Japan.

Egypt, Pakistan, Philippines, Italy, Netherlands and Germany all appear to provide examples of land prices, in relation to rents, quite out of line with any expected rate of return in alternative investments. The quite exceptional figure for Germany appears to be an "expectation" factor of still further government bounties to some of the most generously supported farmers in the world. In some of the Asian cases the multiple may represent a scarcity of alternative opportunities for investment. It is surprising that the multiple is so much lower in India than it is in Pakistan.

[8] *Journal of Economic History*, Dec. 1967, p. 585.

In Indonesia Penny[9] finds land prices averaging 5 times the gross product in Java and $2\frac{1}{2}$ times in Sumatra—perhaps about 10 times the rent in both cases.

Finally we may tabulate the remaining available information on land prices, expressed both in money and wheat equivalents, where no comparative information about rents is available. Even without this information, however, they yield results of some interest.

This additional information shows how wide the range of land prices in terms of wheat equivalent can be, from sparsely populated to densely populated countries.

Some of the individual results are of considerable interest. Perhaps the highest-priced agricultural land in the world is in Barbados. Though subject to fluctuations in sugar prices over the past century, it has always been high. The Governor's letter of 4th July 1676[10] stated that "there is not a foot of land in Barbados that is not employed even to the very seaside ... so that whoever will have land in Barbados must pay dearer for it than for land in England". The high prices in Ethiopia also appear to reflect high population density in this highland region.

An exceptionally low figure for the United States in 1784 was cited by Benjamin Franklin,[11] who was clearly trying to attract settlers. "An hundred acres of fertile soil full of wood may be obtained near the Frontiers in many places for 8 or 10 guineas", he wrote. This corresponds to only 0.03 ton/hectare if he was quoting in sterling, or 0.02 in local currency.

It is known that much of central China was literally depopulated in the bloody Taiping and Nien Wars of the 1850s. Ping Ti Ho states[12] that even by 1953 the central provinces had not recovered their 1850 population. We see an extraordinarily dramatic illustration of what happened in the figures of the fall in land prices.

The Indian series showing the rise in land values in Punjab is of considerable interest. This refers to irrigated land, though not every unit of land can count on irrigation water each season. The Deccan, on the other hand, is unirrigated, rather low-rainfall land. Compared

[9] *Bulletin of Indonesian Economic Studies*, Oct. 1966.
[10] *Calendar of State Papers*, Colonial Series, p. 421.
[11] See *American Review*, Winter 1961.
[12] *Studies on the Population of China 1368–1953*, Harvard University Press.

TABLE 40

	Currency unit	Land prices per hectare	
		In money	In wheat equivalent
AFRICA			
Ethiopia	1969	£	15	22.5
Nigeria	1951	£	17	0.7
AMERICA			
Barbados	1838	£	197	14.7
	1884	£	195	22.6
	1907	£	74	10.6
	1911–13	£	116	14.6
	1941–7	£	320	21.2
	1960–3	£	1770	71.0
Brazil 1st-class land	1935	$U.S.	5	0.2
Brazil 1st-class land	1948	$U.S.	45	0.7
Brazil 2nd-class land	1948	$U.S.	30	0.5
Canada Average	1910	$C	82	2.2
	1925–9	$C	94	2.0
Saskatchewan	1929	$C	62	1.6
	1940	$C	35	1.8
	1954	$C	74	1.5
Jamaica	1959	£	86	3.5
U.S.A.	1780–2	$	2	0.03
	1818	$	5	0.7
	1850	$	27	0.86
	1860	$	40	0.96
	1870	$	45	1.21
	1880	$	47	1.18
	1890	$	53	1.82
	1900	$	49	2.18
	1910	$	98	2.89
	1925	$	132	2.78
	1935	$	78	2.35
	1940	$	78	3.11
	1950	$	160	2.19
	1955	$	211	2.85
	1960	$	288	4.50
ASIA			
China (Taiping War Area)	1850	$C	480	16.3
	1860	$C	12	0.3
Turkestan	1938	$C	28	1.0

TABLE 40 *(cont.)*

		Currency unit	Land prices per hectare	
			In money	In wheat equivalent
ASIA *(cont.)*				
India Punjab	1885	R	74	1.4
	1900	R	190	2.2
	1910	R	306	3.5
	1916	R	560	5.1
Deccan	1914	R	198	2.0
	1921–9	R	277	1.8
	1930–9	R	347	4.2
	1949–52	R	830	2.4
All India	1950	R	1420	4.1
	1961	R	1640	3.5
Median of districts	1957–8	R	1635	3.6
Rajasthan	1957–8	R	250	0.6
Kerala	1957–8	R	4700	10.5
Japan	1950	000 yen	208	3.9
Cultivated land	1951	000 yen	291	4.7
	1952	000 yen	447	6.2
	1953	000 yen	633	7.1
	1954	000 yen	936	11.2
	1955	000 yen	1160	13.6
	1956	000 yen	1357	16.3
	1957	000 yen	1525	17.8
Flat land over 70% cultivated	1966	000 yen	2180	18.3
Semi-flat land 41–70% cultivated	1966	000 yen	1960	16.4
Malaysia	1965	$M	3365	11.5
Pakistan E.	1964	R	10,300	16.0
Thailand	1965	baht	3500	3.5
EUROPE				
Belgium	1846	fr	2630	8.3
	1880	fr	4250	16.5
	1928	fr	28,000	16.1
	1950–2	fr	115,000	25.6
Finland	1935–8	000 marks[a]	137	3.5
	1951–4	Do.[a]	200	5.2
	1954–5	Do.[a]	214	5.5

[a] Of 1951–4 purchasing power.

TABLE 40 *(cont.)*

	Currency unit	Land prices per hectare	
		In money	In wheat equivalent
EUROPE *(cont.)*			
Finland *(cont.)*	1964–5	New mark	2747	3.5
France	1815	fr	840	3.3
	1835–40	fr	1060	4.2
	1929	000 fr	7.55	5.6
	1950	000 fr	127	4.9
	1955	000 fr	147	4.3
	1963	New Fr	4860	9.5
	1967	New Fr	6320	11.7
Germany	1861	Mark	662	2.9
	1889	Mark	2400	12.3
	1901	Mark	2220	12.8
	1912	Mark	1705	8.7
	1953–7	Dm	3690	9.4
	1958–62	Dm	8230	21.2
	1963–7	Dm	8530	22.5
Jugoslavia	1939	000 dinar	25.5	25.5
	1954	000 dinar	166	8.3
Netherlands	1912	g	1640	16.9
Portugal	1965	000 esc	17	8.4
Romania	1886	lei	330	1.3
	1905	lei	535	3.4
	1911–16	lei	988	5.5
	1929	000 lei	32	4.3
Spain	1960	peseta	8000	1.9
Do. irrigated	1960	peseta	40,000	9.5
Sweden	1938	kr	232	1.3
	1962	kr	3500	7.6
OCEANIA				
Australia				
Wheat and fat-lamb area	1937–9	£A	16.1	2.3
Do.	1953–4	£A	44.3	1.7
Good rainfall sheep land	1953–4	£A	36.8	1.5
Queensland wheat–sheep	1952–3	£A	9.0	0.4
Queensland pastoral	1952–3	£A	1.8	0.1
New South Wales N.W. slopes	1960	$A	99	1.9
Wheatland	1967	$A	185	4.0

THE VALUE OF AGRICULTURAL LAND

TABLE 40 *(cont.)*

	Currency unit	Land prices per hectare	
		In money	In wheat equivalent
OCEANIA *(cont.)*			
New Zealand dairyland	1913	£N.Z. 30.0	3.6
	1926	£N.Z. 49.2	3.7
	1938	£N.Z. 38.4	5.7
	1955	£N.Z. 127	5.0
Good rainfall sheep land	1953–4	£N.Z. 13.6	0.6
Canterbury Plains	1959	£N.Z. 62	2.5

Sources and Notes to Table 40

Ethiopia   Oxford University Geography Department Expedition to the Gamu Highlands. The results are quoted in maize (converted at factor of 0.75) for barley land, and the higher figure for "chero" root crop land.

Nigeria   Galletti, Baldwin and Dina, *Nigerian Cocoa Farmer.*

Brazil   Waibol, *Geographical Review*, Oct. 1948 and *Foreign Agriculture* (U.S. Department of Agriculture), June 1945 (for São Paulo).

Canada   Lattimer, *Canadian Political Science Association*, 1951, and Robinson, *Farm Economics*, July 1957 (for Saskatchewan).

Guatemala   Higbee, *Geographical Review*, Apr. 1947.

Jamaica   Edwards, *An Economic Study of Small Farming in Jamaica.*

United States   *Historical Statistics of the United States* for 1850. The 1900 proportion of total real estate represented by buildings was assumed to apply. Commager and Nevins, *The Heritage of America* (Connecticut in 1780–2 and Ohio prairie). From 1850, *Agricultural Finance Review*, June 1963, and *Agricultural Statistics of the U.S.* Inclusive of buildings.

China   1850 and 1860, Richthofen, quoted in Ping Ti Ho, *Studies in the Population of China 1368–1953*, Harvard University Press. Grain prices were not available for this period. In the eighteenth century Chinese prices were rising steadily, but at the end of the century rice cost only 10 silver dollars (7.2 ounces of silver) per ton, much below European prices of that time. It is assumed that Chinese prices went on rising with the impact of world trade, and that in 1850 and 1860 they stood at two-thirds of prices in U.S., which at that time also used a dollar based on silver (i.e. $44/ton in 1850, $53/ton in 1860).

Turkestan   Chang Chih-yi, *Geographical Review*, 1939.

India   Punjab—Calvert, *Indian Journal of Agricultural Economics*, Dec. 1918, prices of cultivated land. Wheat prices from Brij Narain, *Indian Economic Life*, p. 105.

Deccan: Diskalkan, "Re-survey of a Deccan village", *Indian Society of Agricultural Economics*, 1960.

All India: Datta and Pradesh, *Conference on Research in National Income*, New Delhi, 1957, and *Agricultural Situation in India*, Aug. 1965, p. 360 (Reserve Bank estimate).

The Reserve Bank of India's "Rural Credit Follow-up Survey", gives data of land values in 1957–8 per farm, from which land values per hectare can be computed for twelve districts, scattered throughout India.

Japan   For 1873, *Journal of Economic History*, 1947.

Economic Research Institute, Economic Planning Agency, Economic Bulletin No. 9 (1950–7), National Assembly for Agricultural Research, Data 80, 1967 (in Japanese).

Malaysia   Purcal, Ph.D. Thesis, Australian National University. Refers to Wellesley Province.

Pakistan   Farouk and Rahim, *Modernising Subsistence Agriculture* (Dacca University).

Thailand   World Land Use Survey 1966. Saraphi (N.E. Thailand).

Belgium   Combe, *Niveau de Vie et Progrès Technique*; Leroy-Beaulieu, *La Répartition des Richesses* (for 1830 and 1865).

Finland   Suomela Research Institute of Agricultural Economics Publications No. 1 and No. 9 (1958 and 1967). (Valuations made on rye, which is taken as three-quarters wheat equivalent.) Of this value buildings in 1964–5 represented 69 per cent, drainage 6 per cent and land itself only 25 per cent. The value of the land including buildings and drainage was 2.1 years' gross output.

France   Fourastié, *Documents pour l'Histoire des Prix*, and Toutain, *Cahiers d'ISEA*, No. 115 for 1815–1929. Prices based on crude average for the whole of France and raised by a factor of 1.27 to allow for waste lands, etc.

Last two data from Brangeon and Rainelli, Centre de Recherches d'Economie Rurale, Rennes, Report No. 62. Figures for all France divided by 33.8 millions (agricultural and pasture area excluding forests). These authors consider that the previous estimate should be raised by 18 per cent.

Germany   Combe, *Niveau de Vie et Progrès Technique* (to 1912). Subsequent data from *Agrarwirtschaft*, Jan. 1970. Refer to Schleswig-Holstein, a comparatively poor region. Buildings excluded.

Jugoslavia   Vinski, International Association for Research in Income and Wealth 1957 Conference.

Netherlands   *International Review of Agricultural Economics*, 1921, p. 394.

Portugal   Bailey and Ambler, private communication.

Romania   Mitrany, *The Land and the Peasant in Rumania*; Manoilesco, *Weltwirtschaftliches Archiv*, July 1935 (for 1932).

Spain   Fr. Garravilla, Burgos, private communication.

Sweden   Statens Jordbruksnamnd, and Wetterhall, private communications. Value excluding buildings.

Australia   Collins, *Quarterly Review of Agricultural Economics*, Oct. 1954. Keen, *Australian and New Zealand Association for the Advancement of Science*, 1959. Bureau of Agricultural Economics, *The Australian Sheep Industry*, Queensland. N.S.W. Wheatland: Tyler, *Review of Marketing and Agricultural Economics*, Mar. 1964, and J. D. Fahy, Goondiwindi, private communications.

New Zealand   Brian Low, Massey College, private communication (sales standardised to representative farm of 200 acres) for dairy land; for sheepland, Keen (see Australian reference above) and Mason, *Australian Journal of Agricultural Economics*, Dec. 1960.

with other densely populated countries, Indian land values are remarkably low, and show some tendency to fall. Even in the most densely populated areas surveyed by the Reserve Bank the equivalent value was only 10.5 tons/hectare. Areas in Punjab, Bengal and Bihar showed values equivalent to approximately 6 tons/hectare.

Japanese land values, it will be remembered from Table 39, reached their maximum in the 1930s at 22.6 tons/hectare. After the virtual confiscation of land not owner-occupied in 1945, real values were very low in the 1950s, perhaps through fear of a second confiscation. As these fears receded, values again rose rapidly, back to about the previous maximum, where they appear to have stabilised.

The low figure for north-east Thailand indicates a comparatively low population density.

Further information for France supplements Table 39. During the

Napoleonic Wars there was a fall in the real value of land, probably associated with loss of population, but a rapid rise appears to have taken place after 1840. From the 1920s to the 1950s there was a spell of low values, followed by a further marked recovery; likewise in Germany.

The low figure for nineteenth-century Romania is noteworthy.

Wheat production in Australia in the 1960s was unrestricted, export markets were temporarily favourable, and also it was subsidised by a levy on internal consumption. As was to be expected, these favours were translated into land values.

INDEX

113